Power-Efficient Computer Architectures

Recent Advances

Synthesis Lectures on Computer Architecture

Editor
Margaret Martonosi, *Princeton University*

Founding Editor Emeritus
Mark D. Hill, *University of Wisconsin, Madison*

Synthesis Lectures on Computer Architecture publishes 50- to 100-page publications on topics pertaining to the science and art of designing, analyzing, selecting and interconnecting hardware components to create computers that meet functional, performance and cost goals. The scope will largely follow the purview of premier computer architecture conferences, such as ISCA, HPCA, MICRO, and ASPLOS.

The Datacenter as a Computer: An Introduction to the Design of Warehouse-Scale Machines, Second edition
Luiz André Barroso, Jimmy Clidaras, and Urs Hölzle
2013

Shared-Memory Synchronization
Michael L. Scott
2013

Resilient Architecture Design for Voltage Variation
Vijay Janapa Reddi and Meeta Sharma Gupta
2013

Multithreading Architecture
Mario Nemirovsky and Dean M. Tullsen
2013

Performance Analysis and Tuning for General Purpose Graphics Processing Units (GPGPU)
Hyesoon Kim, Richard Vuduc, Sara Baghsorkhi, Jee Choi, and Wen-mei Hwu
2012

Automatic Parallelization: An Overview of Fundamental Compiler Techniques
Samuel P. Midkiff
2012

Phase Change Memory: From Devices to Systems
Moinuddin K. Qureshi, Sudhanva Gurumurthi, and Bipin Rajendran
2011

Multi-Core Cache Hierarchies
Rajeev Balasubramonian, Norman P. Jouppi, and Naveen Muralimanohar
2011

A Primer on Memory Consistency and Cache Coherence
Daniel J. Sorin, Mark D. Hill, and David A. Wood
2011

Dynamic Binary Modification: Tools, Techniques, and Applications
Kim Hazelwood
2011

Quantum Computing for Computer Architects, Second Edition
Tzvetan S. Metodi, Arvin I. Faruque, and Frederic T. Chong
2011

iv

Quantum Computing for Computer Architects
Tzvetan S. Metodi and Frederic T. Chong
2006

Power-Efficient Computer Architectures: Recent Advances

Magnus Själander, Margaret Martonosi, and Stefanos Kaxiras

ISBN: 978-3-031-00617-3   paperback
ISBN: 978-3-031-01745-2   ebook

DOI 10.1007/978-3-031-01745-2

A Publication in the Springer series
SYNTHESIS LECTURES ON ADVANCES IN AUTOMOTIVE TECHNOLOGY

Lecture #30
Series Editor: Margaret Martonosi, *Princeton University*
Founding Editor Emeritus: Mark D. Hill, *University of Wisconsin, Madison*
Series ISSN
Print 1935-3235   Electronic 1935-3243

Power-Efficient Computer Architectures

Recent Advances

Magnus Själander
Uppsala University

Margaret Martonosi
Princeton University

Stefanos Kaxiras
Uppsala University

SYNTHESIS LECTURES ON COMPUTER ARCHITECTURE #30

ABSTRACT

As Moore's Law and Dennard scaling trends have slowed, the challenges of building high-performance computer architectures while maintaining acceptable power efficiency levels have heightened. Over the past ten years, architecture techniques for power efficiency have shifted from primarily focusing on module-level efficiencies, toward more holistic design styles based on parallelism and heterogeneity. This work highlights and synthesizes recent techniques and trends in power-efficient computer architecture.

KEYWORDS

power, architecture, parallelism, heterogeneity

Contents

CHAPTER 1

Introduction

1.1 FROM THE BEGINNING...

Managing the power dissipation of current computer systems is a Grand Challenge problem. Power affects computer systems at all scales: from the computational capacity of our large-scale data centers [18], to the processing performance of our high-end servers [25], and the battery life and performance of our mobile devices [45, 152, 207].

To today's computer architects, the emergence of power as a grand challenge [91] may seem like a relatively recent issue, but the reality is that the very earliest computer systems faced vexing power challenges. For example, the ENIAC computer first became operational in 1946, and its initial press release [191] included this alarming text[1] about its kilowatts of power dissipation:

The ENIAC consumes 150 kilowatts. This power is supplied by a three-phase, regulated, 240-volt, 60-cycle power line. The power consumption may be broken up as follows; 80 kilowatts for heating the tubes 45 kilowatts for generating d.c. voltages, 20 kilowatts for driving the ventilator blower and 5 kilowatts for the auxiliary card machines.

Over the decades that have followed, computer systems benefited from technology refinements that improved circuit performance, cost, and power. Gordon Moore's predictions of technology scaling linked integration levels (transistors per chip) to production cost [138]. For many years, these cost-driven integration improvements also translated quite naturally into performance improvements.

Nearly concurrently, Dennard articulated a scaling principle that would lower supply voltages as transistors became smaller [42]. It is Dennard scaling that enables the transistor increases predicted by Moore's Law to be parlayed into performance improvements and power savings.

Despite the benefits of Dennard Scaling, the power dissipation of integrated systems has spiked before. Most notably, some of the high-performance bipolar ECL processors of the late 1980s and early 1990s dissipated over 100 W [100]. While these designs were impressive and offered then unmatched performance, the costs and challenges of designing and packaging such chips [99] are considered to have played a major role in the adoption of CMOS technology for *high-performance* designs. The challenge for us today is that we have reached a similarly difficult operating point regarding CMOS power dissipation, but without any viable alternative technology available to turn to next.

---

[1]The text's punctuation is as in the original.

Table 1.1: Dennard scaling rules [42]

Device or Circuit Parameter	Scaling Factor
Device dimension T_{ox}, L, W	1/k
Doping concentration Na	k
Voltage V	1/k
Current I	1/k
Capacitance eA/t	1/k
Delay time per circuit VC/I	1/k
Power dissipation per circuit VI	$1/k^2$
Power density VI/A	1

In addition to technology trends, our industry is also driven by application trends that further drive the need for power-efficient computing systems. Compared to 30 years ago, much more of today's computing market (phones, laptops, tablets, games) is at least moderately mobile, and therefore designed with battery life as an important characteristic. In years past, mobile technology would simply "trickle down" from the server world, as old designs are shrunk in new processes. Now the mobile domain is a distinct target with great need for nimble, adaptive power/performance tradeoffs [70]. At the other extreme—the server and enterprise end of the spectrum—power also matters more than ever. Data centers co-locate thousands of high-end servers, and are often limited by their ability to offer sufficient power and cooling to sustain their desired execution throughput [19].

Thus, both technology and application drivers have placed us at a point where power-efficient computation is both vital to future computer systems' viability, and also increasingly difficult to achieve. The following sections elaborate on this.

1.2   THE END OF DENNARD SCALING AND THE SWITCH TO MULTICORES

The power problem as we have faced it over the past decade is largely due to two effects. First, it is primarily a consequence of the end of the Dennard scaling rules (Table 1.1) that parlayed Moore's Law into performance and power benefits for more than three decades. Dennard scaling rests on several key shifts that can be made when transitioning to a smaller feature size. For example, smaller transistors can switch quickly at lower supply voltages, resulting in more power efficient circuits and keeping the power density constant. But supply voltages cannot drop forever. A breakdown of Dennard scaling occurred when voltages dropped low enough to make static power consumption a major issue. Second, even if Dennard scaling had continued on-track, our propensity for faster clock rates and larger die sizes meant that each generation's power dissipation was scaling up faster than Dennard effects were able to hold it in check.

In particular, a key advantage of CMOS technology for many years was its lack of static power dissipation. That is, the complementary p- and n-networks in CMOS gates (theoretically) do not allow any path from supply voltage to ground, consuming power only when switching (dynamic power and some glitch power). Static power consumption was therefore safely ignored at the architectural level. However, when technology scaling broke the 100 nm barrier, transistors showed their analog nature: they are never truly off, and this allows sub-threshold leakage currents to flow. Worse, sub-threshold leakage currents are exponential to threshold voltage reductions. In Dennard scaling, the major mechanism to improve power efficiency is the reduction of the supply voltage, which assumes a reduction of the threshold voltage (since the difference of the two voltages dictates transistor switching speed). The rise of static power brought a complete stop to the power benefits architects took for granted for many technology generations. One-time reductions of static power consumption are possible but the trends remain the same with scaling. For example, current technologies employ multi-gate transistors also known as FinFETs. In these transistors, a fin between the drain and source is "wrapped" by silicon in a non-planar fashion to enable the gate to better encompass the channel, which reduces leakage. As one particular example, Intel switched to 3D or tri-gate transistors in their 22 nm technology [24]. While this change provided a step reduction in leakage going from 32 nm to 22 nm, further reductions will be limited in subsequent scalings.

In the 1980s and early 1990s (the heyday of Moore's Law scaling), architects primarily improved performance by exploiting instruction-level parallelism (ILP)—parallelism found in the dynamic instruction stream during execution of a program. To discover and exploit this parallelism, significant hardware resources were thrown at the problem. Sophisticated techniques such as out-of-order execution, branch prediction and speculative execution, register renaming, memory dependence prediction, among others, were developed for this purpose. These approaches can be highly complex and do not scale well. This results in diminishing performance returns (number of instructions executed in parallel) for increasing hardware investments. Dynamic power scales even worse, deteriorating the power efficiency of such approaches. In fact, as illustrated in Figure 1.1, power dissipation scales as performance raised to the 1.73 power for the typical ILP core: a Pentium 4 is about six times the performance of an i486 at 23 times the power [71]!

The shift to multicore architectures started in 2004 as a reaction to this looming problem of increased power consumption and power density. Effectively, we abandoned frequency scaling (which resulted in significant increases in both dynamic and static power consumption) in favor of laying down more cores on the same chip. This dramatic shift to chip multiprocessors (CMPs) in the past decade is a response to the power wall and the end of Dennard scaling. In particular, Borkar et al. [26] walks through an example for 45 nm technology that is still instructive. For a 45 nm chip with 150M transistors, Figure 1.2 shows a range of possible options for implementing the processor. To abide by the total limit of 150M transistors, one can use more logic transistors (x-axis) in opposition with fewer cache transistors. The resulting power dissipation is shown on the left y-axis, and the resulting cache capacity is shown on the right y-axis. As one increases

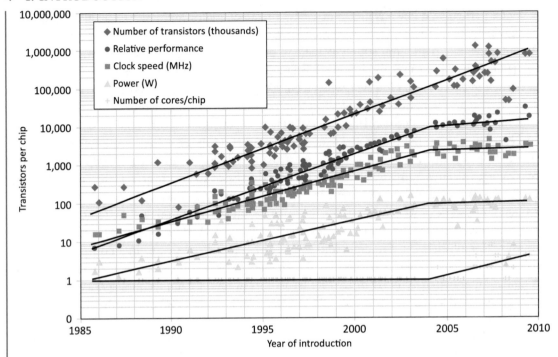

Figure 1.1: Moore's Law and corollaries. Data shows scaling trends, with clear shifts in trend lines at roughly 2004 [37].

the number of transistors devoted to logic, the power dissipation increases (because caches are "cool" from a power standpoint). Pollack's rule [154] argues that microprocessor performance scales roughly as the square root of its complexity, where the logic transistor count is often used as a proxy to quantify complexity.

From these rules of thumb, multiple parallel cores essentially always beat monolithic single cores on power-normalized performance. For example, Figure 1.3 shows three approaches that use parallel cores to enhance throughput while maintaining the same power envelope [26]. Case A (far left of Figure 1.2) harnesses 6-way parallelism at a fairly coarse-grain, and is out-performed by Case B (far right), which is more aggressively parallel, when enough thread/task level parallelism exist in the workload. Case C represents heterogeneous parallelism, in which two large cores are mixed with several small ones, to good effect. An even more heterogeneous approach would be to include some specialized accelerators, which use very few transistors or chip area, but have large performance and power benefits when applicable.

Overall, these examples and rules of thumb begin to explain the direction that industry has taken: a quick and aggressive adoption of medium-scale, on-chip parallelism. Parallelism helps with the impending power wall, by offering a path to high performance that does not rely on

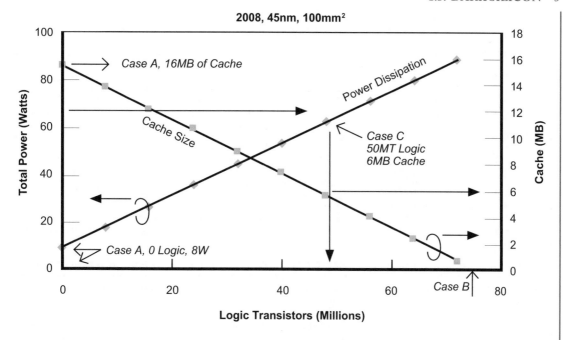

Figure 1.2: A range of implementation options trading off processor area devoted to cache, and resulting power tradeoffs [26].

high clock rates and high supply voltages. Parallelism—particularly heterogeneous parallelism—also helps with the so-called "utilization wall" [186] and the "Dark Silicon" problem [54], as discussed next.

1.3 DARK SILICON, THE UTILIZATION WALL, AND THE RISE OF THE HETEROGENEOUS PARALLELISM

Our inability to scale a single core to further exploit ILP in a power efficient manner turned computer architecture toward exploring alternative kinds of parallelism (task/thread parallelism, data parallelism). Multicore and manycore architectures are designed for explicit parallelism, and recalling Figures 1.2 and 1.3, they offer greater performance-per-watt than large monolithic approaches. Unfortunately, even homogeneous CMPs will not be sufficient to solve the power problem for more than a few more generations [137]. This road is also faced with the same problems as with the single core architecture: we are unable to efficiently extract sufficient speedup from parallel programs (Amdahl's Law [7]).

Furthermore, some postulate a near future in which the number of *dynamically active* transistors on a die may be greatly constrained, forming the "utilization wall" [186]. The concept of the utilization wall is that power envelopes may lead to scenarios in which few (perhaps 20% or less)

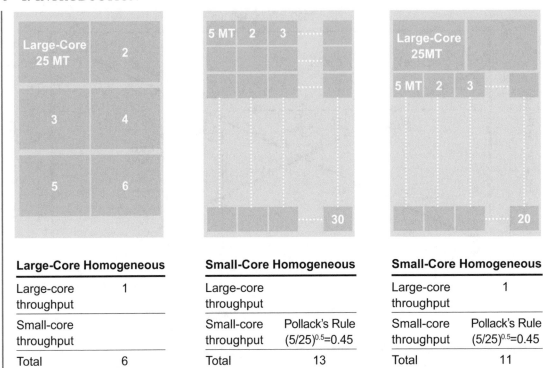

Large-Core Homogeneous		**Small-Core Homogeneous**		**Small-Core Homogeneous**	
Large-core throughput	1	Large-core throughput		Large-core throughput	1
Small-core throughput		Small-core throughput	Pollack's Rule $(5/25)^{0.5}=0.45$	Small-core throughput	Pollack's Rule $(5/25)^{0.5}=0.45$
Total throughput	6	Total throughput	13	Total throughput	11
(a)		(b)		(c)	

Figure 1.3: Enhancing throughput while maintaining power envelope [26].

of a chip's transistors can be "on" at a time. The argument for this possible future is exemplified in Figure 1.4. If transistor density increases in line with Moore's Law, a 45 nm chip will shrink to one-quarter the size at 22 nm in 2014, and one-sixteenth at 11 nm in 2020. Using the ITRS roadmap [91] for scaling, the smaller chips would be more efficient, drawing the same power at 22 nm even though the peak frequency increases by a factor of 1.6, and 40% less at 11 nm with 2.4 peak clock speed. But, if we maintain the same chip area, we can pack four times the number of transistors at 22 nm and 16 times at 11 nm. For the same initial power budget this means that only 25% of the transistors can be powered-up in 22 nm, and 10% in 11 nm. These results are also supported in recent academic studies [54].

The answer to the challenge of the utilization wall is the rise of heterogeneous architectures where some general-purpose cores are augmented by other cores of different microarchitectures or even specialized accelerators that offer outstanding performance-per-watt by being very lean hardware designs for a particular computational purpose. The approach of heterogeneous paral-

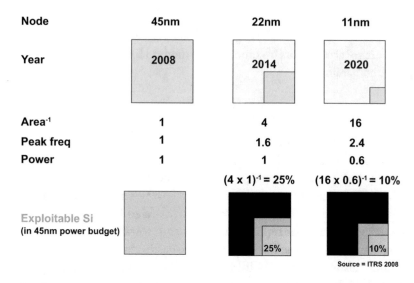

Figure 1.4: A depiction of Dark Silicon trends as seen by ARM [2, 143].

lelism with specialized accelerators is well-suited for "Dark Silicon" scenarios. A large number of accelerators can be built on the same chip to be woken up only when needed. These heterogeneous architectures are fast becoming the dominant paradigm after multicore.

In fact, we do not need to speculate about future heterogeneity, as heterogeneous parallel computing is here today. If we examine the product lines from the major chip manufacturers we see that they now have separate multicore x86 architectures targeted at high performance (2–12 cores, 100W, 100 GFLOPs) and low power (1–2 cores, 10W, 10 GFLOPs), and are integrating data-parallel graphics cores onto their CPU devices with distinct programming and memory models [28]. In the embedded world, there are a range of cores at different performance/efficiency points (1–8 cores, 2W, 10 GMIPS) with a range of programmable graphics cores [133]. NVIDIA, Samsung, and Qualcomm all sell heterogeneous ARM/GPU processors with many fixed-function accelerator blocks for the smart phone market [12, 144, 159], and there are multiple start-ups with 64–100 core devices [183] for networking and telecom. This present-day processor heterogeneity forces system and software designers to address the difficult optimization challenge of choosing the right processor (both at design time and runtime) for their product's power and performance requirements.

Beyond simply considering heterogeneity in the types of instruction-programmable cores on-chip, the field is also increasingly considering approaches involving specialized accelerators that may not be instruction-programmable, and that are tuned to particular application kernels of interest. Specialized accelerators are a particularly natural response to the Dark Silicon scenario in which we may have many more transistors than what we can power up at once. With these "dormant" transistors we could build a plethora of specialized accelerators that cost little either in terms of "active" area or power when not in use. The expectations of generality—all transistors must be useful to all applications—shift considerably in a Dark Silicon world, and what once might have been viewed as "niche" accelerators become a viable method for achieving performance goals under dramatic power constraints.

1.4   OTHER ISSUES AND FUTURE DIRECTIONS

Overall, computer systems have reached an intriguing inflection point. For architects, power has been *a* fundamental design constraint for well over a decade now, with the initial reaction being fairly localized, per-module efforts to improve power efficiency. These efforts have been the equivalent of turning lights off in unused rooms of one's house—very sensible, but insufficient in leverage to dramatically change the overall power-performance design landscape. The second wave in power-aware computing has been the recent and seismic shift toward on-chip parallelism.

Software and Programmability Issues: In many ways, the hardware industry's shift toward parallelism has occurred much faster than the abilities of the software and systems designers to react. We know how to build CMPs, and we must build them to keep Moore's Law rolling along. But we do not yet know how to program them efficiently—both in terms of software development time and in terms of getting the best power-performance outcomes from them. Furthermore, the shift toward on-chip accelerators offers even greater programmability challenges. Finally, there are a host of programmability concerns that emanate from the basic goal of elevating power to a first-class design constraint alongside performance. For example, from a power perspective, information on the relative criticality of different communication or computation operations may be very useful, but current programming models offer few abstractions or constructs to help programmers manage this.

Reliability Tradeoffs: Until now, power-performance tradeoffs have been viewed by architects as a two-dimensional optimization landscape. There is emerging research, however, on the possibilities of *three-dimensional* optimization scenarios in which power, performance, and *reliability* are traded off against each other. Such tradeoffs are already frequently considered at the device and circuit level, but in ways that enable the architecture and software levels to be shielded from their effects; abstraction layers give the impression of perfect reliability even when device or circuit tricks are being employed [52].

Intuitively, there seem to be rich opportunities for raising the abstraction layer at which reliability, energy, and performance are traded off, in order to enable architects to exploit them as well. For example, operating with smaller supply voltage noise margins (by lowering supply voltage) may offer high leverage on power savings, at the expense of possible calculation or storage errors. Likewise, reducing or eliminating parity/checksum protection on memory or interconnect also seems to offer some intuitive power/reliability tradeoff possibility. The key research questions in this space, however, focus on whether the power/performance benefits achievable through some approaches are large enough to be appealing given the serious impact of relaxing reliability guarantees to software.

Beyond the Processor Core: Much of the "first wave" of power optimizations focused on the CPU itself, because the most serious thermal and power density concerns were experienced there. And even more specifically, most optimizations were focused on the CPU's processor cores and cache memories. As we look, however, to future power issues and ideas, there is a growing need to look beyond the processor core. Data communications and on-chip interconnect will play an increasingly important role in power dissipation, especially since the adoption of parallelism has led to much higher levels of data motion and inter-processor communication in many cases. One also needs to consider the energy issues related to the memory hierarchy as well. Chapter 4 covers these topics in this book, but considerable future work in this area is likely to be forthcoming.

1.5   ABOUT THE BOOK

To conclude this introduction, we include here some further explanations about the book that may be helpful to readers in finding relevant material and in comparing with contents from a prior Synthesis Lecture by Kaxiras and Martonosi [103].

1.5.1   DIFFERENCES FROM THE PRIOR SYNTHESIS LECTURE [103]

We view the two books as largely complementary. The first book offered extensive details on the sorts of local, per-module power optimizations that comprised the industry's "first wave" response to the power challenge. In this current book, we take a more holistic view. As a result, both the structure and content of the book have changed dramatically.

There are three core chapters, which synthesize highlights of power-efficient computer architecture techniques. Chapter 2 covers voltage and frequency scaling issues, with a particular emphasis on trends and techniques that have emerged in the years since the first edition of the book. Chapter 3 covers techniques related to specialization and heterogeneity that have emerged with greater prominence in the five years since Dark Silicon began to emerge. Chapter 4 covers the power implications of data motion and storage, again with a particular emphasis on more recent techniques and trends. Finally, Chapter 5 concludes the book.

We note that while the power dissipation of main memory has emerged as an important problem, we feel that this topic is too broad to be covered well as part of this book. Thus, this book does not cover main memory issues in earnest, and we hope that another synthesis lecture will take on this topic in detail.

Finally, a note on power modeling approaches. These were covered in the first book [103]. While new tools and modeling environments have been created in the years since then (e.g., [76, 122, 127]), these tools employ fairly similar basic philosophies and approaches as previous generations of tools. Thus due to space and scope constraints we have chosen not to cover them further here.

1.5.2   TARGET AUDIENCE

This book is written for researchers who have taken a basic course in computer architecture, and are interested in becoming somewhat fluent in the power implications of architectural choices. We envision it being particularly useful for a new graduate student who may be familiar with the basics of computer design and architectural simulations, but perhaps has been less exposed to power issues. In addition, systems researchers from related fields (e.g., operating systems, compilers, parallel programming, and others) may find the book useful for understanding some of the architectural viewpoints and issues interposed between the technology challenges emerging "from below" and the applications trends "from above."

CHAPTER 2

Voltage and Frequency Management

Issues addressing dynamic power have predominated the power-aware architecture landscape. Amongst these dynamic power techniques, most methods focus on dynamic voltage and frequency scaling (DVFS). The intuition behind many of these approaches [82, 92, 93, 164, 197, 200, 201] is that if the processor and memory operate largely asynchronously from each other, then the processor can be dialed down to much lower clock frequencies during memory-bound regions, with considerable energy savings but no significant performance loss. A good overview of early DVFS techniques is given by S. Kaxiras and M. Martonosi [103]. This chapter discusses the motivation for these techniques, DVFS technology trends, DVFS power and performance models, and OS managed DVFS. Also discussed, is the concept of criticality in parallel applications and its applicability for controlling DVFS.

2.1 TECHNOLOGY BACKGROUND AND TRENDS

The basic dynamic power equation: $P = CV^2Af$ clearly shows the significant leverage possible by adjusting voltage (V) and frequency (f) [32, 78]. If we can reduce voltage by some small factor, we can reduce power by the square of that factor. Reducing supply voltage, however, might possibly reduce the performance of systems as well. Reducing supply voltage often slows transistors such that reducing the clock frequency is also required.

The benefit of DVFS is that within a given system, scaling supply voltage down offers the potential of a cubic reduction in power dissipation. The downside is that it may also degrade performance. If the program runs at lower power dissipation levels, but for longer durations, then the benefit in terms of total energy will not be cubic. It is interesting to note that while voltage/frequency scaling improves the energy delay product (EDP) (because the reduction in power outpaces the reduction in performance), it can do no better than break even on the energy delay-squared product (ED^{2}P) metric—assuming performance is proportional to frequency.

However, there is the opportunity for more benefit when the relationship between performance and frequency (f) is *non-linear*. This happens when there is *slack* in the execution, which cushions the effect of frequency scaling. This is the main premise of the DVFS approaches that we discuss in this chapter.

2.1.1   RELATION OF V AND f

Figure 2.1 shows the relation between supply voltage, frequency, leakage, and energy efficiency. The darker curve in the left graph of Figure 2.1 illustrates how the maximum operational frequency depends on the supply voltage, while the lighter curve illustrates how the total power depends on the same. The darker curve in the right graph of Figure 2.1 illustrates how the energy efficiency changes with the supply voltage. The greatest energy efficiency is achieved just above the subthreshold voltage at which point the leakage power is much lower than if the supply voltage is raised, as shown by the lighter curve.

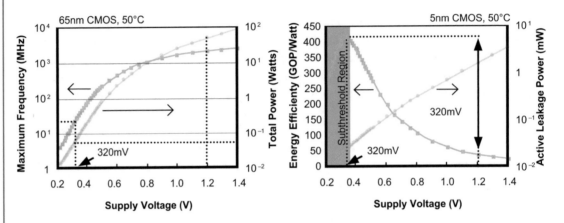

Figure 2.1: Showing the dependency between (left graph) maximum frequency and supply voltage and (right graph) leakage and energy efficiency and supply voltage [26].

The frequency (f) at which a circuit can operate is proportional to the difference in supply voltage (V) and subthreshold voltage (V_{th}) as given by $f \propto \frac{(V - V_{th})}{V}$. To maintain the same maximum operating frequency any scaling of the supply voltage requires an equal scaling of the subthreshold voltage. But since subthreshold leakage increases exponentially with the reduction in subthreshold voltage, static power has become an increasing concern for submicron technologies.

The increase of subthreshold and gate leakage in submicron technologies, the two major contributors to static power, has led to a stagnation in the scaling of the subthreshold voltage. Over the years, this led to the shrinking of the supply-voltage range for which dynamic scaling can be performed. The maximum operational voltage has continued to shrink while the minimum operational voltage has remained roughly the same for the past decade, which is illustrated in Figure 2.2.

While for many technology generations, the voltage and frequency range enabled us to successfully reduce energy by adapting to current performance requirements of the system, the end of Dennard scaling reduces the opportunity for using DVFS in contemporary systems.

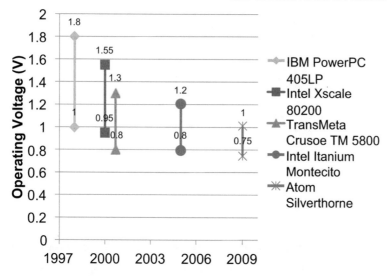

Figure 2.2: Shrinking V range [192]. IBM PowerPC 405LP: 1.0V–1.8V, TransMeta Crusoe TM 5800: 0.8V–1.3V, Intel XScale 80200: 0.95V–1.55V, Intel Itanium Montecito: 1.2V–0.8V, Intel Atom Silverthorne: 1.0V–0.75V.

2.1.2   TECHNOLOGY SOLUTIONS

The increase in leakage currents, and thus static power, is the fundamental cause for the DFVS problems we are facing in future technologies. Thus, solutions aimed to combat leakage at the technology or other levels are of particular interest here.

The issue with increasing leakage currents comes from the challenge of controlling the conducting channel between the drain and source of a transistor. When the channel length is of the same order of magnitude as the source and drain depletion regions, which it is for submicron technologies, a number of physical properties affect the capability to turn the channel off. These physical properties have been termed short-channel effects [147].

There are several ways to mitigate these effects:

- To counteract short-channel effects the gate oxide thickness can be reduced. This provides better control of the channel without increasing the subthreshold voltage.

- Thinner gate oxides increase the chance for gate leakage currents where electrons tunnel through the oxide. To reduce gate leakage currents new materials with High-i[0] dielectric structures are used instead of the more conventional silicon dioxide.

- Multi-V_{th} designs where performance-critical parts are implemented using low-V_{th} transistors that are fast but leaky, and less critical parts are implemented using high-V_{th} transistors that are slower but much less leaky.

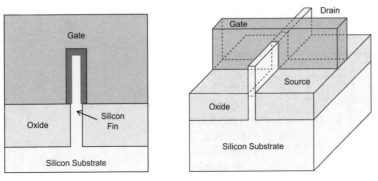

Figure 2.3: Intel's 22 nm FinFET tri-gate transistors [24].

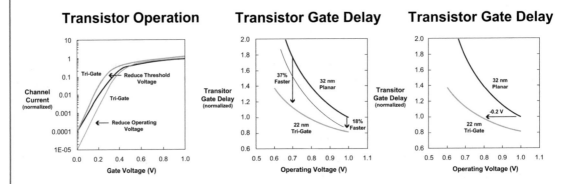

Figure 2.4: Intel's 22 nm FinFET characteristics [24].

- Another option is to construct three-dimensional gates where the channel is surrounded on three sides by the gate, so-called FinFET [75] transistors. For example, Intel's 3D Tri-Gate transistors in 22 nm depicted in Figure 2.3 form conducting channels on three sides of a vertical fin structure, which provides "fully depleted" operation. The "fully depleted" characteristics of such transistors provide a steep sub-threshold slope that can be used to lower leakage. The steep sub-threshold slope (Figure 2.4, left graph) enables the use of a lower threshold voltage, in turn enabling the transistors to operate at lower voltage (100–200mV) to reduce power and/or improve switching speed (Figure 2.4, left graph). Compared to a planar FET, a FinFET transistor can operate faster at the same operating voltage (Figure 2.4, middle graph), or at lower operating voltage at the same speed (Figure 2.4, right graph).

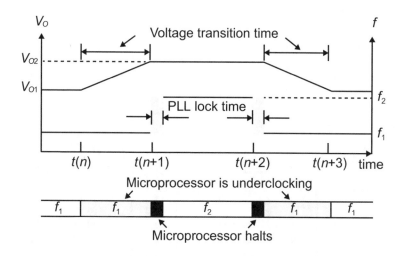

Figure 2.5: Voltage stepping vs. frequency stepping latencies [150].

2.1.3   DVFS LATENCY

Some of the implementation characteristics for DVFS can have significant influence on the strategies an architect might choose, and the likely payoffs they might offer. For example, what is the delay required to engage a new setting of (V, f)? (And, can the processor continue to execute during the transition from one (V, f) pair to another?) If the delay is very short, then simple reactive techniques may offer high payoff. If the delay is quite long, however, then techniques based on more intelligent or offline analysis might make more sense.

Typically, DVFS transition latencies are in the order of tens of microseconds [150]. This limits the timeframe for DVFS techniques to that of operating system (OS) task-switching, at best. The reason for such a high transition latency is the latency of voltage stepping by the off-chip voltage regulator (VR). Typical off-chip voltage regulators constructed of discrete components (board-level inductors and capacitors) operate at a switching frequency of less than 5 MHz and have relatively slow voltage adjustment capabilities [110] (see Figure 2.5). Frequency stepping is faster, accounting for only 7.6% of the delay overhead, depending on the frequency steps and the response time of the PLL [150].

A solution to the voltage-stepping latency problem is to use *on-chip* voltage regulators operating at much higher switching frequencies (see Figure 2.6). This solution is first explored at the architectural level in the work of Kim et al. [110]. The downside of using on-chip voltage regulators is that conversion efficiency drops, leading to increased energy losses [110]. However, the benefits from fine-grain (V, f) control outweigh the negatives [110]. Intel's Haswell processor family has a fully integrated on-chip voltage regulator [88].

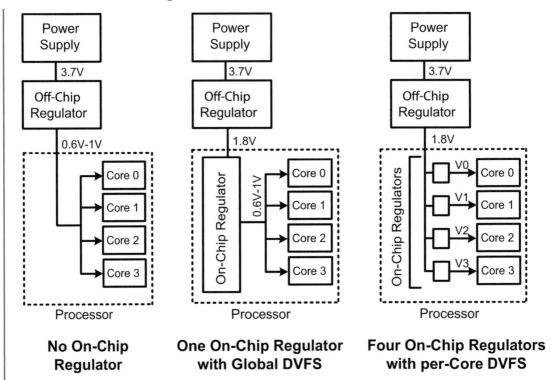

Figure 2.6: On-chip voltage regulators [105].

2.1.4   DVFS GRANULARITY

The bulk of the DVFS research has focused on cases in which the entire processor core operates at the same (V, f) setting but is asynchronous to the "outside" world, such as caches. In such scenarios, the main goal of DVFS is to capitalize on cases in which the processor's workload is heavily memory-bound. In these cases, the processor is often stalled, waiting on memory, so reducing its supply voltage and clock frequency will reduce power and energy without having significant impact on performance.

Some of the early architectural work on DVFS focused on opportunities within multiple-clock-domain (MCD) processors [168, 169, 181, 198, 199]. The rationale for MCD processors is that as feature sizes get smaller, it becomes more difficult and expensive to distribute a global clock signal with low skew throughout the whole processor die.

The multi-core paradigm, is a natural MCD design [44]. Cores can be scaled more aggressively than memories, resulting in multiple voltage and clock domains with asynchronous communication between them. One example of this aggressive scaling is *core boosting* where the supply voltage of a core is temporarily increased if power and temperature envelops are not breached in order to boost the performance of individual cores. The power budget cannot sustain permanent core boosting of all the cores at the same time and a boosted core can result in reduced power budgets for the remaining cores that have to operate at a lower frequency. Both Intel [86] and AMD [3] implement core boosting in their latest processor architectures.

Another related question regards whether continuous settings of (V, f) pairs are possible, or whether these values can only be changed in fixed, discrete steps. If only discrete step-wise adjustments of (V, f) are possible, then the optimization space becomes difficult to navigate because it is "non-convex." As a result, simple online techniques might have difficulty finding global optima, and more complicated or offline analysis becomes warranted. Industry has now settled in a discrete and small number of DVFS states.

2.2   MODELS OF FREQUENCY VS. PERFORMANCE AND POWER

A naive approach for estimating the impact of frequency scaling in performance is to assume that the execution time of an application is inversely proportional to the processor's frequency. This, however, is a very pessimistic estimation of the penalty implied by frequency scaling. The processor communicates with the memory in an asynchronous way, i.e., accessing the main memory is not affected by the processor's frequency. Therefore, changing the core's frequency only affects the parts of the application that do not exhibit any accesses to the main memory. In other words, memory-bound programs can be scaled down in frequency with only minor impact on their performance. This effect has been exploited *empirically* in a number of approaches [82, 92, 93, 164, 197, 200, 201]. Recently, analytical models quantify the relationship of performance to f scaling, giving a better insight for such techniques. As with other DVFS techniques applied at the OS level the key enabler is *slack* [103]—in this case, the slack that results from waiting for memory (*memory slack*). We present these models here, for the insight they offer on DVFS behavior.

2.2.1   ANALYTICAL MODELS

Karkhanis and Smith introduced a first-order *interval-based* analytical model for superscalar processors [102]. Eyerman and Eeckhout together with Karkhanis and Smith later refined this model and proposed practical approaches for the on-line collection of its parameters [57]. The interval model estimates performance as a function of various event counts, such as cache misses, branch mispredictions, etc., and the instruction-level parallelism (ILP) that is inherent to an application. Apart from predicting raw performance numbers, an equally important contribution of the interval model is to serve as a framework for understanding how the so-called "miss events" affect the performance of a processor. In this direction, three different groups, working independently (Keramidas et al. [106], Eyerman and Eeckhout [56], and Rountree et al. [160]) extended the basic interval model to describe how frequency scaling affects the execution time of a program.

The model introduced by Karkhanis and Smith breaks the execution of a program into intervals. During "steady state" intervals, the execution rate in instructions-per-cycle (IPC) of the processor is only limited by the processor's width and the program's dependencies within an instruction window (for sufficiently large instruction windows). Steady state intervals are interrupted by "miss-intervals," which introduce stalls cycles to the processor. A miss-interval starts with a miss-event (off-chip load accesses in our case), and lasts until the pipeline reaches again a steady state (a period related to the memory latency). The realization that underlies all DVFS models is that *core frequency scaling in these models is equivalent to changing the memory latency in cycles*. This is depicted in the diagram of Figure 2.7.

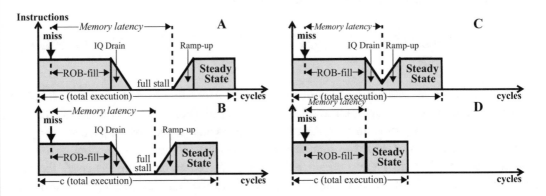

Figure 2.7: Performance scaling with f. Memory latency, *measured in cycles*, shrinks as we scale f lower, eliminating the stalls caused by the miss [106].

Scaling down frequency does not always result in a proportional slowdown in the processor's performance. This is due to the asynchronous way that a processor core communicates with the remaining system and main memory. Processor core operations that do not incur an L1 cache miss have a fixed latency in terms of cycles because every core component and the L1 cache operate under the same clock signal. Thus, in a program with few L1 misses, scaling the clock frequency does not change the total number of cycles required for the program to execute. Of course, the duration of the cycle (cycle period) is affected in an inverse proportional way, thus the execution time is also inversely proportional to frequency. However, latencies to components outside the core and the L1 cache scale proportional to the difference between the core and the components' clock frequencies. This effect is particularly tangible for off-chip accesses (e.g., to main memory) as they operate at a relatively low frequency and have long latencies.

Figure 2.8 shows in more detail what happens upon an access that misses in the last level cache (LLC). When the miss occurs, the processor continues to issue instructions at a steady-state rate, since there are still instructions that do not depend on the pending miss. This area is called *ROB-fill* (reorder buffer fill), and corresponds to the number of cycles required until the pending miss reaches the head of the reorder buffer. At this point, no more instructions can be

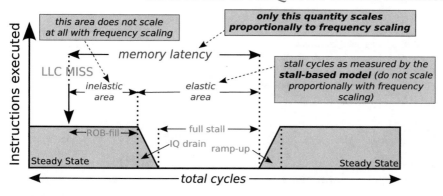

Figure 2.8: Interval model for f scaling [106].

committed, thus no more new instructions can enter the instruction window. The processor now has a limited pool of instructions to choose from for issuing to the execution units—a pool that gradually shrinks as instructions execute. This results in the IQ (instruction queue) drain area of the graph: the processor has fewer and fewer candidate instructions to pick from, thus the issue rate starts dropping. At some point, all of the remaining instructions will depend on the pending LLC miss, and the processor stalls waiting for the miss to be resolved (full-stall area). When the miss is finally resolved, pending instructions can execute, new instructions can enter the instruction window and, thus, the issue rate ramps-up to the stead-state rate again.

Examining Figure 2.8, and keeping in mind that time is measured in processor core cycles, the models identify how different regions are affected by frequency scaling. Individual regions of the miss-interval are affected by frequency scaling in different ways. Regarding the ROB-fill area, the number of cycles that it takes until the reorder buffer fills up *does not depend on frequency*. All of the operations that occur during this period are core operations, synchronized at the same clock, and changing the clock-period does not affect the number of cycles these operations take. IQ drain, full-stall and ramp-up intervals, on the other hand, scale with frequency in a way that the total memory latency (measured in processor cycles) will scale proportionally to frequency.

There are two models that can be constructed based on these observations, a simple model called *stall-based*, which is practical for on-line implementations, and a more elaborate, more accurate *leading-loads* model, which, however, is less practical given the current event-counter support in commercial processors. Both are explained below.

Stall-based Model

The simplest model assumes that the time it takes to fill the ROB is negligible and therefore the bulk of the memory latency is elastic to frequency scaling. This corresponds to the case of an isolated LLC-miss. When multiple LLC-misses overlap (Figure 2.9), the issue rate begins to rise after the first miss has been serviced but drops again when the second miss reaches the ROB

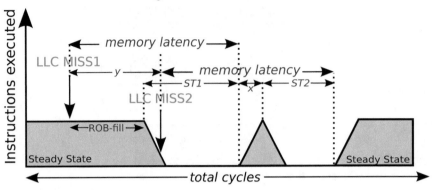

Figure 2.9: Interval model for multiple overlapping misses [106].

head. When the data of the second miss arrive from the main memory, the issue rate rises until it reaches the steady-state rate. In Figure 2.9, although the second miss occurs y cycles after the occurrence of the first miss the second miss reaches the ROB head x cycles (where $x \ll y$) after the first miss has been serviced. As a result, two separate stall intervals appear: ST1 and ST2.

The stall-based model assumes that either stalls are generated by an isolated miss or the sum of stalls generated by overlapping misses (e.g., ST1 and ST2) are both approximately equal to the memory latency (in cycles). Because memory latency is proportional to core frequency, these quantities are approximately proportional to frequency as well. Consequently, the total number of stall cycles is approximately proportional to frequency, while the total number of non-stall cycles (steady state) is independent of frequency (measured in cycles).

The advantage of this model is that only in-core information (number of stall cycles) is used to predict performance under various frequencies. However, the model assumes that ROB-fill is negligible, which is a source of errors especially in applications characterized by little dependence between instructions and thus large ROB-fill time.

Leading Loads Model

While the stall-based model assumes that the number of stall cycles is approximately equivalent to the memory latency for both isolated and overlapping misses, the "leading loads" model recognizes that there is an inelastic area in the miss-intervals (the ROB-fill time) that does not scale with frequency (in cycles). While this is not modeled or measured directly, its implications, especially with respect to overlapping misses, are captured in the model: stall cycles of only the first miss of a group of overlapping misses or the stall cycles of an isolated miss are taken into account.

Leading load models require an accurate accounting of *the number of groups of overlapping misses*. Support for measuring these at run-time is not readily available in contemporary processors but has been proposed by Eyerman and Eeckhout [56].

There are several assumptions underlying these models, for example, the conditions that stall the processor [56], and the effect of store misses [106]. Perhaps one of the most important simplifications is that the access latency of the main memory is considered constant, even though in reality it may vary considerably. This assumption is addressed in the work of Miftakhutdinov et al. [134] where the authors show improvement in accuracy by adopting a more elaborate memory latency model.

2.2.2   CORRELATION-BASED POWER MODELS

Power models can be split in two broad categories: highly parameterized models that try to estimate power consumption of non-existent hardware (CACTI [119], WATTCH [27], Mc-PAT [127]) and models that characterize a specific hardware. This subsection briefly discusses the second approach.

The standard method for building a model to characterize the power behavior of a specific system is to train a correlation model [39, 64, 65]. Different benchmarks (or microbenchmarks) are used to stress different parts of the system (caches, ALU, out-of-order engine, etc.) and performance counters are programmed to measure different statistics such as caches misses, branch mispredictions, and integer/floating point instructions. At the same time, power consumption is measured for each of the benchmarks of the training set, and curve-fitting techniques are used to derive the relationship between the performance-counter statistics measured and the power observed for each benchmark. The mathematical representation of the above is to express power as a linear function of all the statistics that are believed to correlate strongly with power consumption,

$$Power = \sum_{k=0}^{i} \frac{param_k \times event_k}{time} \tag{2.1}$$

Power and values for the performance counter events are used to derive the parameters $param_k$ that best fit the observed measurements. Notice that this model is tightly connected with a specific (V, f) pair, the one that the benchmarks were run at while creating the correlation model. Recently, Spiliopoulos et al. [177] proposed an extension to this approach to create a unified power model that estimates power consumption not only for the voltage and frequency that were used during the training of the correlation model, but for any frequency of interest. Their model acknowledges that, since power consumption is given by the formula $P = P_{static} + af C V^2 = P_{static} + f C_{eff} V^2$, and C_{eff} is the only non-deterministic part of the equation, the correlation formula can now be constructed for C_{eff}

$$C_{eff} = \sum_{k=0}^{i} \frac{param_k \times event_k}{cycles} \tag{2.2}$$

The parameters are derived in a similar way, but the model is now decoupled from specific voltage and frequency values—the accounting of V and f is done once using the generic equation

$P = P_{static} + f C_{eff} V^2$. The most important aspect of this approach is that, by using the analytical DVFS models presented in the previous subsection, statistics can be collected in a base frequency and estimated for a target frequency, therefore power can be estimated not only for the base but also for the target frequency.

2.2.3   A COMBINED POWER AND PERFORMANCE MODEL

Throughout this chapter, it is clear that abstract models play a significant role toward understanding the efficiency of DVFS. It is also important to realize that DVFS is something that should be explored not statically for each program, since the impact of (V, f) scaling into performance and energy dissipation can vary significantly within the same program, depending on the behavior of the different phases. Spiliopoulos et al. built a tool that takes all this into account and analyzes the behavior of different phases in terms of energy dissipation and execution time for any frequency of interest [178]. This tool, called PowerSleuth, combines three basic components:

- The first component is a phase-detection library called ScarPhase [167]. ScarPhase is an online phase detection tool that is capable of detecting phases at a very low overhead (2%) concurrently with the execution of the application.

- The second component is an approximation of the stall-based model discussed earlier in this chapter. The "approximation" is due to the fact that contemporary processors do not provide the necessary information for applying the model as it is, hence certain approximations have to be taken in order to predict the impact of frequency scaling in the execution time of an application.

- The final component is a model to correlate the power consumption to various performance events in a real processor, similarly to other works [39, 64, 65].

PowerSleuth can capture the runtime variation of an application over different phases, as well as estimate how execution time and energy are impacted by frequency scaling.

2.3   OS-MANAGED DVFS TECHNIQUES

Architectural techniques for dynamic voltage and frequency scaling first appeared in the literature pertaining to the system (or operating system) level. Commercial implementations controlled at this level are also the most common form of DVFS (e.g., Intel's Enhanced SpeedStep® [85] and AMD's PowerNow!™ [4]).

Weiser et al. first published on this type of DVFS [193]. They observed that *idle time represents energy waste*. To understand why this is, consider the case of a processor finishing up all its tasks well within the time of a scheduling quantum. The remaining time until the end of the quantum is *idle time*. Typically, an idle loop is running in this time, but let us assume that the processor can be stopped and enter a sleep mode during this time. One could surmise that a profitable policy would be to go as fast as possible, finish up all the work and then enter the sleep mode for the idle time and expend little or no energy. But that is not so.

2.3.1   DISCOVERING AND EXPLOITING DEADLINES

Whereas the DVFS techniques of Weiser et al. are based on the idle time as seen by the operating system (OS) (e.g., the idle loop), Flautner, Reinhardt, and Mudge look into a more general problem of how to reduce frequency and voltage without missing deadlines [59]. Their technique targets general purpose systems that run interactive workloads.

What do "deadlines" mean in this context? In the area of *real-time* systems, the notion of a deadline is well defined. Hard real-time systems have fixed, *known* deadlines that have to be respected at all times. Since most real-time systems are embedded systems with a well-understood workload, they can be designed (scheduled) to operate at an optimal frequency and voltage, consuming minimum energy while meeting all deadlines. An example would be a mobile handset running voice codecs. If the real-time workload is *not* mixed with non-real-time applications, then DVFS controlled by an on-line policy is probably not necessary—scheduling can be determined off-line.

Flautner et al. consider an entirely different class of machines. In general-purpose machines running an operating system such as Linux, program deadlines have more to do with user perception than with some strict formulation. Thus, the goal in their work is to discover "deadlines" in irregular and multiprogrammed workloads that ensure the quality of interactive performance.

The approach to derive deadlines is by examining communication patterns from within the OS kernel. Application interaction with the OS kernel reveals the so-called *execution episodes* corresponding to different communication patterns. This enables the classification of tasks into interactive, periodic producer, and periodic consumer. Depending on the classification of each task, deadlines are established for their execution episodes. In particular, the execution episodes of interactive tasks are assigned deadlines corresponding to the user-perception threshold, which is in the range of 50–100 ms. Periodic producer and consumer tasks are assigned deadlines corresponding to their periodicity. All this happens within the kernel without requiring modification of the applications.

2.3.2   LINUX DVFS GOVERNORS

DVFS is handled in the Linux kernel by the so-called *cpufreq* component. Cpufreq is modular, in the sense that it separates the low-level operations for setting voltage and frequency and the higher-level operations of selecting the frequency to run at. The low-level operation are handled by the *cpufreq driver*, which is responsible of setting up the voltage and frequency tables, exposing this information to the higher levels, and controlling voltage and frequency according to the commands that it receives from the higher levels of the software stack. On the other hand, the *frequency governors* are responsible for selecting the frequency to run at depending on the workload executing on the system. Essentially, a single *cpufreq driver* is enough for a specific system, but various different *frequency governors* can co-exist in the system, and the user can easily switch between them at runtime depending on the power and performance profile that better suit the current use-case. Here we briefly discuss the standard governors that come with every Linux distribution:

- **Performance:** The simplest possible governor, runs always at maximum frequency.

- **Powersave:** The complete opposite of the *performance governor*, runs always at minimum frequency. This governor minimizes the power consumption but may severely prolong the execution time of certain applications.

- **Ondemand:** The default governor used in many systems, tries to set frequency based on the CPU utilization. Frequency is reduced in steps when utilization drops below some certain low-threshold, and it is restored back to the maximum value if CPU utilization rises above the high-threshold. Notice that CPU-utilization is expressed at a system-level, which means that a CPU that is waiting for some cache misses to be serviced by the main-memory still appears to be fully utilized to the OS.

- **Conservative:** Very similar to the *ondemand* governor, with the difference that this governor is much more conservative to increase the frequency when an increase in the CPU-utilization is detected.

- **Interactive:** This is also a utilization-based governor, but it does not use any threshold to rise and drop the frequency. Instead, the governor uses the observed CPU-utilization to determine what fraction of the maximum frequency should the frequency be.

Green Governors

Based on the analytical performance-model and the power-correlation model presented earlier in this chapter, Spiliopoulos et al. proposed a class of frequency governors that intelligently select the voltage and frequency setting that optimizes an energy efficiency metric [177]. Being based on models that estimate performance and power as a function of frequency, these governors can be tuned for various different policies, i.e., for optimizing different power efficiency metrics. The authors demonstrate two different policies, optimizing energy delay product (EDP), and also a variation of this policy that minimizes EDP to the extent possible, without harming performance by more than 10% compared to the execution of the application at maximum frequency. However, any policy that involves some tradeoff between power and performance can be created by using the two models presented in the previous subsections.

Green Governors are implemented as Linux kernel modules, in a similar way as the standard frequency scaling governors that come with all the popular Linux distributions. Given the performance and power models presented in previous subsections, the idea is really simple. The kernel profiles the workloads running on the system for a short period of time, typically a few tens of milliseconds, and gathers the information required by the models. The models determine what the performance and the power consumption of the workload would have been in any other possible frequency, and based on the policy used, the optimal frequency for the interval that just executed is estimated. Assuming the next interval has a similar behavior with the previous one, the best frequency estimated for the previous interval is the frequency that the upcoming interval will execute at.

2.4   PARALLELISM AND CRITICALITY

As chip multiprocessors (CMPs) have become the predominant general-purpose, high-performance microprocessor platform, it becomes important to consider how DVFS management can be applied to them most effectively. One major design decision concerns whether to apply DVFS at the chip level or at the per-core level. As with other MCD designs, per-core DVFS is considered more expensive; it requires more than one power/clock domain per chip, and synchronizer circuits are required to avoid metastability between domains. On the other hand, multiple clock domains may be employed anyway for circuit design or reliability reasons, in addition to voltage and frequency control.

Early research has explored the benefits of per-core versus per-chip DVFS for CMPs. For example, on a four-core CMP in which DVFS was being employed to avoid thermal emergencies (rather than simply to save power), a per-core approach had $2.5\times$ better throughput than a per-chip approach [44]. This is because the per-chip approach must scale down the entire chip's (V, f) when even a single core is close to overheat. With per-core control, only the core with a hot spot must scale (V, f) downward; other cores can maintain high speed unless they themselves encounter thermal problems.

The discussion so far in this chapter centered on the DVFS behavior of a core running a single-threaded program. When considering one single-threaded application in isolation, one need only consider the possible asynchrony between compute and memory.

While a multicore processor can be used to run independent programs for throughput, its promise for single-program performance lies in thread-level parallelism. Managing power in a multicore when running parallel (multi-threaded) programs is currently a highly active area of research. Many research groups are tackling the problem, considering both symmetric architectures, which replicate the same core and asymmetric architectures that feature a variety of cores with different power/performance characteristics [117]. Independent DVFS for each core [8], a mixture of chip-wide DVFS and core allocation [123], or work-steering strategies at the program level in heterogeneous architectures [117, 139] are considered. In this scenario, reducing the clock frequency of one thread may impact other dependent threads that are waiting for a result to be produced. Thus, when considering DVFS for parallel applications, some notion of critical path analysis may be helpful.

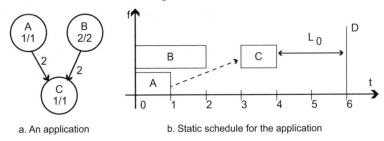

a. An application   b. Static schedule for the application

Figure 2.10: An application and its static schedule [136].

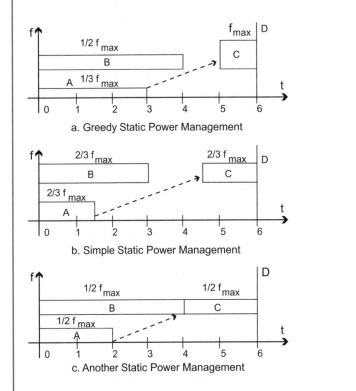

a. Greedy Static Power Management

b. Simple Static Power Management

c. Another Static Power Management

Figure 2.11: Various static schedules that result from distributing the global slack and scaling (V, f) [136].

2.4.1   THREAD- AND TASK-LEVEL CRITICALITY: STATIC SCHEDULING

Mishra et al. performed some of the early work (2003) on power management of multithreaded workloads [136]. They examined the case of distributed real-time systems where a set of real-time tasks with precedence constraints and deadlines is statically scheduled to execute. Their starting point is a static schedule generated by any list scheduling algorithm, such as for example the one given in Figure 2.10. In this example three tasks, A, B, and C, run for the specified amount of time (at a given f). The communication between A and C (with a cost of two time units) is indicated with a dotted line in the schedule. A global deadline D is set at time 6.

Prior to their work, simple static algorithms distribute the global slack (L_0 in Figure 2.10) proportionally to the tasks and by scaling (V, f) obtain energy savings respecting at the same time the precedence constraints, the communication, and the deadlines. Scheduling results of three static algorithms are shown in Figure 2.11. The authors argue that these static approaches are not optimal for parallel and distributed systems when *parallelism varies in an application*. Thus their proposal is based on the intuition that more energy savings can be obtained by distributing proportionally more slack to sections with higher degree of parallelism. The effect is to further reduce the idle periods in the system. Although their approach is static and relies on a statically scheduled real-time task graph, it saw the seeds for later work. The authors keenly observe:

"Due to synchronization of tasks and parallelism of an application, gaps may exist in the middle of a static schedule. After distributing global static slack, gaps in the middle of the schedule can be further explored. Finding an optimal usage of such gaps seems to be a non-trivial problem. One simple scheme is to stretch tasks adjacent to the gap when such stretching does not affect the application timing constraints."

Subsequent work in this area addresses exactly this observation: how to detect "gaps" in execution and exploit them for energy savings. The key is to detect load imbalance between threads and slow down non-critical threads to achieve energy savings. However, it has proven more fruitful to detect *critical* threads that cannot be slowed down (on account of their direct impact on the total program execution) than to try to assess which threads are non-critical.

2.4.2 THREAD- AND TASK-LEVEL CRITICALITY: DYNAMIC SCHEDULING

The first attempt to exploit "gaps" in execution of parallel programs is the proposal for *Thrifty Barriers* by Li, Martinez, and Huang [124]. The gaps in question are experienced by threads that arrive early at a barrier. At this point a thread can select to set its core in a sleep mode. However, because entering the sleep mode involves significant latency, and exiting the sleep mode late can impact performance (as it can delay crossing the barrier for all threads) it is important to be able to gauge the size of the "gaps" before sleep decisions are made. To solve this problem Li, Martinez, and Huang use prediction. The size of the gaps cannot be accurately correlated to the executing threads, since it is unpredictable how the barrier load imbalance is distributed between the threads. However, they found that the interval between successive invocations of the same barrier is quite stable and therefore predictable. Thus, in their proposal, each thread arriving at a barrier has a prediction of the inter-barrier interval and can gauge its own compute time. The difference between the two is the predicted gap for this particular thread's execution. This enables threads to decide whether to set their core to sleep or not while waiting for the end of the barrier.

If a thread decides to enter a sleep mode it must also wake up in a timely manner so as to not penalize the whole execution. A thread sets an alarm (according to its predicted gap), which is also guarded by a safety net in case the prediction is widely off. The safety net is provided by the coherence mechanisms and concerns the flag that is typically used in barriers to signify the crossing of the barrier by the last thread. This flag's address is monitored by the cache controller of the thread's sleeping core, and in case an external coherence invalidation is detected for this address the core is immediately woken up.

Whereas Li, Martinez, and Huang aimed at detecting and predicting the gaps in execution, Cai et al. first turned the attention to *thread criticality* [30]. They proposed a technique to dynamically detect critical threads in a parallel region. As a critical thread they simply define the thread with the longest completion time in the parallel region. To find a solution Cai et al. constrain the problem. They detect imbalance in parallel loop iterations. For this reason, the approach is not appropriate for dynamic scheduling of irregular code but rather tuned to static scheduling as offered, for example, in OpenMP.

The approach is simple: a *meeting point* is established (by the programmer or compiler, or conceivably by the hardware itself) such that each thread regularly visits this point. For example, a meeting point can be the back edge of a parallel loop. The number of times that each thread passes the meeting point corresponds to the amount of "work" the thread executes. A thread running behind others in terms of the number of times it passes the meeting point, means it is critical (i.e., has the potential of delaying the whole execution as it does not progress as fast as the others on its assigned work). The slack between a slow thread and other threads can be estimated by the difference in the times they pass the meeting point.

Knowing the relative rate at which threads go through their assigned work permits slowing-down the fast threads (since the total execution time is bounded by the slowest) to gain in energy or alternatively speeding up the slow (critical) thread to gain in performance. In both cases the DVFS decisions are taken locally by each core running a thread. This, however, assumes global knowledge at every core. The authors propose that each core periodically broadcasts to all other cores, the number of times it passed a meeting point. Once this global knowledge is established then distributed decisions can be made. Each core attempts to match the rate of the slowest core by proportionally adjusting its frequency.

2.4.3   CRITICALITY

While the work of Cai et al. focused specifically on parallel loop iterations and monitored each core's relative progress, the general problem of how to assess thread criticality in arbitrary code remained unsolved. A first solution came by turning to prediction. Bhattacharjee and Martonosi [22] proposed to predict—rather than to try to accurately track—thread criticality.

Their goal is to construct predictors out of readily available run-time information from hardware performance counters. To this end they examined which metrics correlate best with thread criticality. Instruction rate (number of instructions executed in a specific period of time) proved to be successful only in some applications, but failed to correlate with criticality in others. However, cache hierarchy events such as misses proved to be much better indicators of criticality across a wide range of applications. A thread that misses in its L1 more than its peers is very likely to be the bottleneck in the computation. The inclusion of L2 misses further improves accuracy for the cases where the L1 misses alone do not provide enough. Establishing a strong correlation between memory system events and criticality, Bhattacharjee and Martonosi are able to construct simple and effective predictors out of information readily available in processors. Similar to other works, once the critical thread is identified, it can be sped up for performance, or the non-critical threads slowed down for energy savings.

Finally, Du Bois et al. take a different view on criticality [48]. They define as critical threads the threads that make others wait via synchronization. Furthermore they define a new metric for criticality that combines both the amount of time a thread executes useful work *and* the number of co-running threads that are waiting—the more threads that are waiting the bigger the impact of managing a critical thread. This refers to the work of Mishra et al. [136], discussed previously, which introduced the notion of distributing slack according to the degree of parallelism.

Du Bois et al. also propose a direct measurement of criticality—rather than predicting it through indirect measurements. For this they propose some modest hardware (counters and logic) to track the time threads are performing useful work, or are waiting. This measurement setup leads to a representation of thread criticality as a "stack," ranging from 0% to 100%, where each thread is ranked according to the percentage it occupies in the stack. The thread that occupies the largest percentage is the most critical, and so on. Management of threads takes the two standard forms: critical thread acceleration for performance and energy optimization by slowing down the least critical of threads. Because of the ranking of all threads and the accuracy of the approach due to direct measurement their results are the most promising yet.

2.5 CHAPTER SUMMARY

DVFS has been the staple of power/energy management. Due to the Dennard-scaling breakdown the useful voltage range has been shrinking. This has turned DVFS to DFS making it harder to achieve efficiency improvements (e.g., EDP). This has forced architects to bet more on the non-linear relationship of performance to frequency and the existence of slack in execution (either memory-system slack or slack due to parallelism). While race-to-halt techniques maximizing frequency to reduce leakage are important, better models of program behavior and better methods of exploiting criticality enable DVFS to remain an important and powerful energy optimization.

CHAPTER 3

Heterogeneity and Specialization

The prior chapter primarily focused on means for achieving power efficiency by providing low-level mechanisms like DVFS that can then be invoked or managed by software. Here, we shift toward higher-level systems design principles that achieve power efficiency more holistically.

In recent years, as power constraints have become more difficult to manage through simple module-level controls, an increased focus has been placed on design styles that can offer better power solutions. The first response has been homogeneous on-chip parallelism, already heavily discussed. While this offers compelling power/performance opportunities, one can still go further. In particular, opportunities exist to differentiate different on-chip processing modules in order to offer a "portfolio" of options with different power/performance characteristics. These can include:

1. Heterogeneous on-chip parallelism in which cores of the same ISA are available on-chip, with different configurations or microarchitectures.

2. Heterogeneous on-chip parallelism in which cores with *different* ISAs are available.

3. Homogeneous or heterogeneous parallelism as above, further augmented by specialized hardware accelerators for particular computations.

4. Heterogeneous designs that exploit *reliability* vs. energy tradeoffs, in addition to performance tradeoffs.

The techniques discussed in the following sections follow the progression of bulleted items listed above. In addition, we note that for each case, there are often opportunities for employing Dark Silicon techniques, to keep some of the processors or modules entirely turned off. In fact, some of the ideas we discuss below were particularly motivated by Dark Silicon arguments, although most have more general power/performance benefits as well. We therefore start with a brief discussion of Dark Silicon, before elaborating on each of the topics above.

3.1 DARK SILICON

Early works in power-aware architecture were motivated by attacks on the classic dynamic power equation $P = CV^2Af$. While some reduced C by reducing effective wirelength, or used V, f management techniques, others focused on reducing the average activity factor, A. Reductions in *average* activity factor can improve average power or aggregate energy usage, but they do not address the *maximum* power problem.

In 2009, the term "Dark Silicon" came into use. It describes a scenario in which fabrication technology can build dense devices with high transistor counts, but power constraints preclude all or many of them switching at once. One of the earliest references to Dark Silicon is ARM CTO Mike Muller's keynote talk at the annual ARM technical conference [132]. Muller foresaw that by 2019, an 11 nm process technology would be able to deliver devices with 16× the transistor count of 2009 parts, but with power budgets so constrained that the fraction of active transistors at any time might be limited to roughly 9 percent.

Responses to Dark Silicon trends come in several forms. Silicon on insulator (SOI) fabrication techniques are increasingly being used to reduce leakage energy at smaller feature sizes [46]. In addition, 3D integrated circuit techniques (3D ICs) also offer promise for their ability to improve the time and energy devoted to on-chip data motion.

At the architecture level, the response to Dark Silicon concerns has also been quite varied, and the response to these trends has led to interesting debate. There is not yet a solid consensus on how designs will adapt to a power-limited rather than transistor-limited future. Possible approaches fall into several categories, from lower transistor counts at one end of the spectrum, to extreme specialization at the other. We discuss some of the ongoing research below.

3.1.1   DARK SILICON ANALYSIS AND MODELS

Esmaeilzadeh et al. discuss the trends leading to Dark Silicon, and provide both analytical models as well as architectural prognostications regarding future paths [54]. In particular, they combine three main modeling techniques: (i) a device-scaling model for area, power, and frequency analysis down to 8 nm; (ii) a core-scaling model to analyze maximum single-core performance and the resulting power dissipation, and (iii) multi-core scaling models that consider different architectural possibilities including homogeneous and heterogeneous CMP approaches. Using these models and studying a range of PARSEC applications [23], they conclude that at 8 nm, over 50% of the chip will be "dark" at any time. Further, they claim that neither CPU-like nor GPU-like multicores will offer sufficient power-performance efficiency to track hoped-for performance scaling at achievable power limits. Papers like this one point the way toward increased use of specialized accelerators or other non-instruction-programmable compute units on future chips.

3.1.2   DESIGNING FOR DARK SILICON: BRIEF EXAMPLES

Some work has proposed architectures directly with the Dark Silicon era in mind. For example, the work on GreenDroid [66–68] and Conservation Cores [186] specifically discusses chip architectures aimed at the Dark Silicon problem. In Conservation Cores, the authors target the irregular integer codes not well served by traditional accelerators. The processor cores, known as *c-cores* use reconfiguration to adapt their execution to particular codes, with the primary target

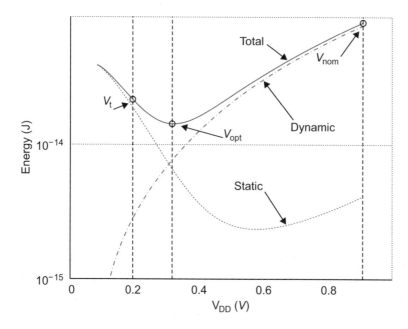

Figure 3.1: Total, static, and dynamic energy across V_{DD} for a 32-nm process [153].

being energy savings rather than performance improvement. They claim whole-program energy improvements by 2.1 for their approach.

The GreenDroid project employs the conservation cores approach to optimize key portions of Google's Android smartphone platform. The GreenDroid chip design is a tiled architecture with 16 tiles. Each of these tiles includes a MIPS CPU, a modest data cache, 6–10 Android c-cores, and routing to interconnect these. C-cores can be automatically generated through their toolflow. In their approach, much of the energy savings (over 10 for some kernels) comes from tailored execution that greatly reduces instruction fetch and decode energy, along with the energy costs of operand data motion.

3.1.3   THE SENTIMENTS AGAINST DARK SILICON

While much of the architecture community is focused on accelerators and specialization as a fundamental response to Dark Silicon trends, there are varying degrees of conviction regarding the urgency and effectiveness of such techniques. Some in industry worry that chip area is too expensive (and large die areas are too influential in lowering chip yields) for silicon to be designed to be (often) dark.

Some argue instead for "dim silicon" where a larger fraction of the silicon is kept powered but at a lower performance [182]. Near threshold voltage (NTV) processing is one of the motivations for dim silicon. By operating close to (but not below) the threshold voltage an optimal tradeoff between dynamic and static energy can be achieved; see Figure 3.1. It would therefore seem beneficial to employ more dim cores, each with a lower performance, to meet the computational demand than a few cores operating at a higher performance. NTV processing is sensitive to process variations though, as small variations in the threshold voltage have significant effect on the performance at this operational point. Furthermore, the workload has to be parallelizable to take advantage of the additional cores, and parallelization overheads need to be accounted for.

Core boosting, e.g., Intel's Turbo Boost [86] and AMD's Turbo Core [3] technologies, can be viewed as one form of dim silicon. In these technologies, the majority of the cores are operated at a low frequency and voltage while selected cores can be boosted to a higher frequency and voltage to execute sequential or critical code faster.

3.2   HETEROGENEITY IN ON-CHIP CPUS

While the term "heterogeneous parallelism" is broad enough to encompass many options, a relatively simple form of heterogeneity is to provide multiple CPUs on the same chip whose *architecture* is identical but whose *implementation* or *configuration* varies. For example, this might include several cores who execute the same ISA, but at different clock frequency settings or with different microarchitectural implementations. The simplicity of such an approach makes it particularly compelling. The first study that showcased the capabilities of a single-ISA heterogeneous architecture is by Kumar et al. [116].

Figure 3.2 shows the increased throughput, for the same chip area, of heterogeneous architectures composed of two successive generations of ALPHA ISA cores over architectures composed of a single type of cores. Taking the configuration with three EV6 and three EV5 cores (3EV6-3EV5) as an example it has roughly the same throughput as five EV6 cores (5EV6) but uses only 70% of the area.

3.2.1   CURRENT INDUSTRY APPROACHES

While Kumar et al. focus on performance, an example of such "single-ISA heterogeneity" focused on power efficiency is ARM's big.LITTLE approach [70]. A simple form of the big.LITTLE architecture pairs a single "big" core (e.g., Cortex A-15 [10], Figure 3.3) with a single "LITTLE" core (e.g., Cortex A-7 [11], Figure 3.3). Both cores are *architecturally* compatible, but they differ in their microarchitectural implementation, which leads to differences in performance and energy. Some computations will have performance requirements that mandate the use of the big core, while other computations may gain sufficient performance from the more energy-efficient LITTLE core (Figure 3.4). The cache and communication subsystems (Figure 3.5) are designed to support fast migration that enables applications to quickly and smoothly switch which core they are using (without, for example, losing their cache state). In addition, both cores can be used simultaneously in multi-processor mode. While we have described this here as a single big.LITTLE pair, it is of course also possible to consider chips that comprise several of these

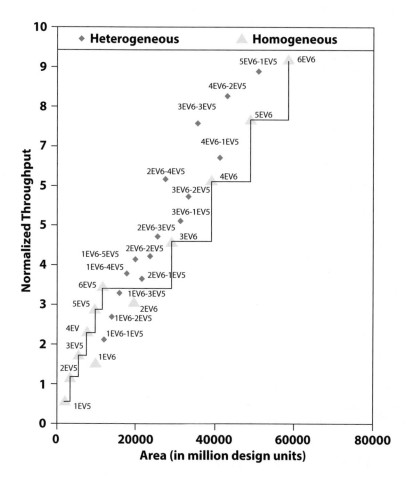

Figure 3.2: Example design space options for Single-ISA Heterogeneity [116].

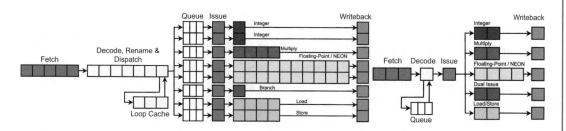

Figure 3.3: ARM Cortex-A15 "big" and Cortex-A7 "LITTLE" pipelines [70].

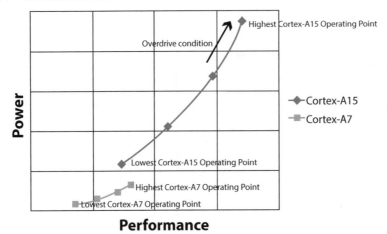

Figure 3.4: ARM big.LITTLE power/performance tradeoff [70].

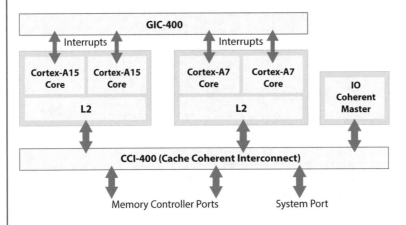

Figure 3.5: ARM big.LITTLE system architecture [70].

pairs, or indeed arbitrary mixes of *N* big with *M* LITTLE cores. Samsung recently introduced a dual-quad version of the big.LITTLE approach, with four pairs of on-chip processors [174].

Another example of single-ISA heterogeneity is NVIDIA's Tegra3 and Tegra4 CPUs [144, 145]. The Tegra3 chip primarily computes using an ARM Cortex-A9 quad-core CPU, but also includes a simpler, low-power, low-clock rate core to which execution can be quickly migrated when the performance requirements are modest or the power constraints are tight. Tegra4 is a similar quadcore-plus-one approach, this time using the Cortex-A15 as its main high-performance workhorse.

3.2.2   RESEARCH AND FUTURE TRENDS

With single-ISA heterogeneous parallelism seeing strong industry adoption (particularly for the mobile sector) researchers have also turned to exploring related software and management issues. For example, Craeynest et al. evaluate scheduling techniques for such chip implementations [184]. Heterogeneity increases the scheduling challenges as migration costs also need to be accounted for to move an application (or thread) from one type of core to another.

In summary, providing heterogeneous microarchitectures unified by a single ISA is appealing as a step toward power efficiency that can extend well beyond DVFS (Chapter 2) but with the relative implementation ease offered by a unified ISA. In fact, approaches like ARM's big.LITTLE can be exploited as different options within the p-states abstractions typically underpinned by DVFS. That is, p-states corresponding to slower, energy-efficient execution are executed on the LITTLE core, while p-states earmarked for high performance are executed on the big core.

3.3   SINGLE-ISA *CONFIGURABLE* HETEROGENEITY

While ARM's big.LITTLE power/performance curve is discontinuous and non-linear, ideally we would prefer *energy proportionality*: pay proportionally (in power or energy) for what you get (in performance). The case for energy proportionality was put forth by Barroso and Hölzle considering servers that rarely reach zero or peak utilization [18].

With respect to multicores, Watanabe, Davis, and Wood propose a configurable architecture called WiDGET (Wisconsin Decoupled Grid Execution Tiles) [192] aiming for power proportionality. Figure 3.6 shows how the architecture scales almost linearly in performance and power compared to two discrete points in the lower left (red circle) and upper right corner (green square) that represent a low-power, low-performance Intel Atom-class processor and a high-performance, high-power-consumption Intel Xeon-class processor respectively. WiDGET consists of a set of instruction engines (IEs) and a sea of execution units (EUs). Cores are dynamically scaled up (or down), depending on application requirements, by coupling more (or fewer) EUs to each IE. Each EU function as an in-order-issue pipeline while each IE consists of front-end and back-end pipeline functionalities comparable to conventional out-of-order pipelines. The IE schedules dependent instructions onto the same EU. This enables independent instructions to execute in parallel on separate EUs and results in instruction-level and memory-level parallelism.

Composable lightweight processors (CLPs) [109] use an explicit data graph execution (EDGE) ISA to schedule an application/thread onto a dynamically configured number of identical simple, narrow-issue processor cores (Figure 3.7). An EDGE ISA reduces the amount of control decisions performed in hardware by grouping dependent instructions into blocks and by providing explicit intra-block dataflow semantics [29]. Within a block, the result of an instruction is directly forwarded to consuming instructions instead of writing them back to a register file. A block can thus be computed in a dataflow manner. This simplifies the composition of simple cores

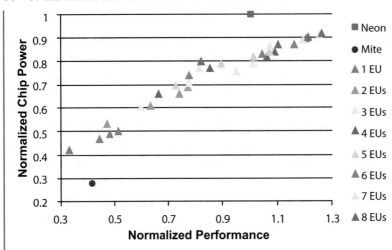

Figure 3.6: WiDGET techniques for power proportionality [192].

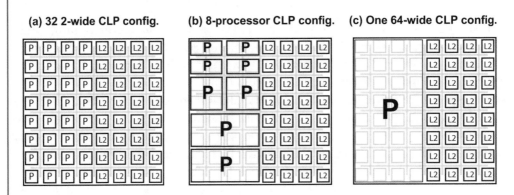

Figure 3.7: Composable lightweight processor (CLP) in three different configurations [109].

into a larger, more powerful core without the need for monolithic structures that spans across all cores, e.g., broadcast busses.

Ipek et al. propose to fuse up to four neighboring cores into one larger CPU when the application demands additional performance [90]. Each core is relatively simple and independent from all other cores, which enables a high level of parallelization. A fetch management unit (FMU) and steering management unit (SMU) coordinates the fetch and steering activities to be performed for a fused core. The FMU manages fetch-related information such as invalidation of TLBs of the separate cores and facilitates branch prediction and subroutine calls. The SMU has a global steering table, free-lists for the register files and rename maps of the fused cores, and steering/renaming logic. The task of the SMU is to steer dependent instructions to the same core for improved performance. An application can request a core fusion or split through two specialized FUSE and SPLIT instructions. On a FUSE commit the following instructions and instruction caches are flushed, and the FMU, SMU, and instruction caches are reconfigured. On a SPLIT commit all in-flight instructions are drained and all register state is moved to core zero of the fused cores before reconfiguring the FMU, SMU, and instruction caches.

3.4   MIXING GPUS AND CPUS

Section 3.2's and 3.3's single-ISA heterogeneity is appealing as a smooth but effective extension from prior DVFS or configuration techniques. Natural further extensions are to more aggressively exploit heterogeneity by dropping the single-ISA requirement. While many mixed-ISA multicore options are possible, one especially noteworthy combination is to combine CPUs and graphics processing units (GPUs) on the same chip.

3.4.1   CPU-GPU PAIRS: THE POWER-PERFORMANCE RATIONALE

GPUs were first introduced primarily for particular applications (e.g., gaming and graphics) with very high and regular parallelism. With power dissipations commonly at 200W or higher [146], the early instinct was not to consider them "power efficient." On the other hand, in terms of energy per calculation, a case can be made that well-tuned, memory-friendly applications can be made very power friendly through the use of GPUs, because their regimented massively parallel single instruction multiple threads (SIMT) execution style dramatically reduces the per-calculation overhead required for instruction fetch and decode. Keckler [104] and Dally [41] postulate that GPUs represent an opportunity to continue to scale performance while abiding by power limits, because GPUs have the potential to reduce energy-per-operation by 10× or more, thanks to such efficiencies. In fact, the Green500 list [69], which tracks the 500 most power-efficient supercomputers, is often dominated by supercomputers that include GPUs. GPUs are not as well-suited to execute irregular code, however, and so combinations of CPU and GPU on the same chip represent a form of heterogeneous parallelism with appealing potential for improving power efficiency.

3.4.2   INDUSTRY EXAMPLES

CPU-GPU combinations are receiving increasing attention both from researchers and as commercial products. AMD and Intel both have major stakes in this area, with products like Haswell, Kavier, and others [6, 87] available as of this writing. Haswell combines a Core i7 or i5 CPU with integrated graphics, and has a specific focus on power-efficient high-performance for laptops and the mid-range mobile market. These integrate technology improvements like the Tri-Gate 3D transistors ("FinFETs") discussed in Chapter 2.1.2 with architecture and system responses for power as well. Likewise, AMD has several products that combine an aggressive CPU (e.g., Jaguar) with GPU technology [5]. General purpose computing on the GPU (GPGPU) is becoming increasingly common for mobile systems. ARM's Mali-T600 GPU family [135] and Imagination

Technologies' PowerVR Series6XT GPU family [83] support general purpose computing and are commonly implemented together with one or more ARM cores on the same system on chip (SoC). The significant commercial presence in this space motivates further research on both hardware extensions beyond these initial pairings, and software and system management for them.

3.4.3 SELECTED RESEARCH EXAMPLES

Some work has laid the groundwork by characterizing GPU power consumption and considering methods for employing DVFS on the GPU side of things [1]. Other work has considered methods by which a chip's aggregate power budget (a shared resource) can be best apportioned between the CPU and GPU as a program executes [190].

Interestingly, the presence of an on-chip GPU has been shown to shift the workload seen by the CPU side of the pair [13]. This in turn shifts design choices regarding what CPU design features make sense from a power or performance perspective. For example, if the high-parallelism portions of code are executed by the GPU, then the CPU sees a relative workload shift toward more irregular and less parallel codes. A recent study [13] notes that this points toward fewer cores on the CPU, but it could also warrant other changes as well.

3.5 ACCELERATORS

Another response to increasing power concerns has been increasing openness to the use of special-purpose on-chip hardware accelerators. Accelerators are motivated by two power sub-trends. First, because they eschew the generality of instruction-level programmability, hardware accelerators can offer excellent performance-per-watt for the computations they support; this is because much of the power overhead of generality—instruction fetch, instruction decode, operand fetch—can be replaced by custom wiring and computation. Second, accelerators play a prominent role in the Dark Silicon story. Previously, a downside to specialized accelerators was a concern about their specificity: namely that they would occupy chip area, but not see broad use. Moving forward, if one envisions a world in which few (~10%) transistors can be switching at any time, then the specificity of accelerators becomes tolerable because they offer good performance-per-watt when in use, and when not in use, the argument is that some of the silicon would be dark anyway.

3.5.1 BACKGROUND

Many research communities have focused considerable effort on designing specialized hardware units for streaming, encryption, and other applications [38, 49]. In addition, industry's embrace of accelerators for strategic applications is also apparent. For example, Apple's iPad tablet includes a **check type** accelerator specifically to support the streamed movies that are so commonly viewed on such platforms. Such work will not be the main focus of our discussion here. Rather, we will focus mainly on the *architectural* research on supporting accelerators and exploiting their power benefits.

3.5.2 SELECTED RESEARCH

For example, one line of work has been to explore how to architecturally integrate the use of accelerators using virtualization and storage sharing. In the Accelerator Store proposal [130], the authors first characterize a suite of potential accelerator units from the standpoint of their storage and area requirements. This characterization leads to an important observation: typical accelerator designs devote between 40% and 90% of their area to generic SRAM storage, used for operands, buffering, and results. Leveraging this observation, the Accelerator Store proposal provides a pool of virtualized SRAM storage resource that can be used by nearby specialized accelerators. In this way, much of the area cost of a specialized accelerator can be amortized over several accelerators instead of just one.

Subsequent work on *Shrink-Fit Accelerators* has built on the Accelerator Store concept [129]. Here, because the accelerator store is a virtualized resource (an accelerator unit can be allocated more or less of it in a fairly seamless manner), the authors demonstrate how to build accelerators that, while specialized, can still smoothly tradeoff different degrees of performance depending on the amount of virtualized resources they are offered. The authors find that, for example, the IDCT accelerator's size can be reduced by 16× (by relying instead on the accelerator store) with small performance and area overheads. Approaches like Shrink-Fit strike a balance offering both the specialization and power efficiency of hardware accelerators, as well as the virtualization and abstraction of more general-purpose environments.

Finally, recent work applies a similar philosophy of balancing generality and specificity by providing a hardware accelerator for a domain specific query language called LINQ [35]. By accelerating LINQ queries, Chung, David, and Lee's accelerator approach, called LINQits, offers some degree of generality since many different uses of LINQ queries are possible. On the other hand, the approach still benefits from the hardware design leanness of a specialized accelerator. The authors prototype their approach on a ZYNQ processor that includes two ARM A9 CPUs with a pool of FPGA resources. From their physical measurements, LINQits improves energy efficiency by 8.9 to 30.6× and performance by 10.7 to 38.1× compared to conventional optimized parallel code running on ARM A9 processors.

3.5.3 INDUSTRY EXAMPLES

Microsoft recently developed techniques to accelerate datacenter workloads by employing field programmable gate arrays (FPGAs) [158]. They argue that most datacenter workloads evolve too quickly to make it economically viable to implement custom accelerators as application-specific integrated circuits (ASICs). FPGAs enable reconfiguration of the hardware and thus enable accelerated parts to be updated as the algorithms progress. It also enables the same hardware to be used for multiple purposes as many different accelerators can be implemented on the FPGAs. The FPGAs can then be configured with an appropriate set of accelerators based on the service that is currently running on the system. A reconfigurable FPGA fabric was developed and evaluated by accelerating the Bing web search engine running on 1,632 servers. By accelerating parts of the ranking algorithm the throughput of the whole system doubled. Put differently, for the same throughput roughly half the number of servers was needed, with significant energy savings as the result.

Intel has announced a new product that combines a Xeon processor with a coherent FPGA in a single package [89]. By coupling the FPGA tightly with the Xeon they hope to improve the performance by $2\times$ thanks to the low-latency, coherent interface.

3.6   RELIABILITY VS. ENERGY TRADEOFFS

As this book has made clear, prolonging the lifetime of Moore's Law and Dennard scaling has required increasingly aggressive approaches for satisfying application performance goals on constrained energy budgets. Until now, we have viewed power and performance options in terms of a two-dimensional Pareto space of optimization tradeoffs.

As power challenges become more extreme, one additional approach to addressing them is to shift toward a design stance in which tradeoffs of circuit reliability are also considered. In essence, the two-dimensional space of performance/energy tradeoffs becomes a three-dimensional Pareto space of performance/energy/reliability tradeoffs.

In current and future designs, chip multiprocessors are likely to be heterogeneous collections of processor cores with different energy, performance, and reliability characteristics. This heterogeneity might arise by design or might arise through technology variations that result in heterogeneous processor behaviors.

Sacrificing reliability for performance may be acceptable if low-level errors can be corrected [52], or if they do not always impact functional correctness. For example, some gate-level errors are not necessarily visible at the architecture or software level, e.g., single-bit errors in a branch predictor or bit-flips in portions of the data cache that are currently unused [142]. Further, many current software applications can also tolerate infrequent low-level errors. For example, image processing and audio applications can both tolerate a modest number of bit-flips in their picture or sound data without any detectable effect on the application behavior. To exploit such data error tolerance, several methods have been proposed either in hardware or at the hardware software interface [14, 53, 55, 73, 101, 121, 163, 185].

Some techniques use language-level constructs to express software error tolerance at the instruction granularity [163] and then schedule onto functional units accordingly. Such techniques require completely reliable instruction-level control flow, with the potential for reliability/energy tradeoffs being limited to arithmetic units and the like. Other techniques constrain execution by eliminating instruction-level control and structuring computations as data streaming through error-susceptible signal processing accelerators [73]. Moving toward increased flexibility, the ERSA project proposes to have one reliable core and N cheaper but less-reliable ones [121]; error-tolerant application phases are written as DOALL parallelism that unfurls across the less-reliable cores, while the primary (reliable) core can enforce timeout limits to ensure appropriate control flow sequencing. Finally, some error-tolerant applications seek to exploit core-to-core heterogeneity in reliability at an even coarser grain. For example, given a "menu" of heterogeneous cores available, one might schedule tasks onto cores such that the resulting schedule is predicted to satisfy reliability and performance constraints with the least energy [202].

Related to error tolerance is the notion of Approximate Computing. Here, the application expresses quality-of-result requirements, and then compiler, system, and hardware techniques may be able to identify and use reduced precision approaches that abide by the quality-of-result requirements while possibly lowering power dissipation. The distinction between approximation and error tolerance lies in the fact that approximate computing techniques can assume much greater system control over the frequency and characteristics by which precision reduction occurs. (With error-tolerance computing, an error model characterizes what sort of errors might occur and when, but ultimately the appearance of errors is stochastic and uncontrolled.) Language-level techniques [163] have potential for harnessing approximate computing opportunities. Hardware-level techniques also seek to harness the power-saving potential of limited precision by using analog computation [179].

3.7   CHAPTER SUMMARY

Perhaps more so than any other sub-topic in Computer Architecture, research in the area of specialization and heterogeneity is moving quickly and offers great potential for power-performance impact over the coming years. While this chapter has highlighted research and industry techniques and trends of particular note, no such synthesis can easily keep pace with a topic area seeing such broad attention. This fast-moving field will offer many further innovations over the coming years.

CHAPTER 4

Communication and Memory Systems

In today's systems, processors remain a leading concern in terms of *power density* but many other components of systems are major power dissipators, worthy of their own attention. Thus, in this chapter, we consider non-computational power issues, such as data movement and data storage. Since entire books have been written focusing solely on interconnect [96] and on memory [94], we do not seek to replicate that material, but instead will select and present aspects of these systems that are particularly relevant from a power perspective.

4.1 THE ENERGY COST OF DATA MOTION: A HOLISTIC VIEW

In current computer systems, the actual arithmetic units are barely visible in a die photo, and the power they dissipate is a small fraction of overall chip power. For example, Steve Keckler and William J. Dally, both from NVIDIA, each gave keynote presentations in 2009 and 2011 that featured the example data illustrated in Figure 4.1 [40, 104].

Figure 4.1 shows a die and illustrates the area and energy dissipated by a 16-bit multiply accumulate (MAC) unit, a 64-bit floating point unit (FPU), on-chip channels of various widths and lengths, and off-chip channels. It shows that a floating point operation is an order of magnitude more energy efficient than moving a word across the die over a distance of half the die length (to access the last level cache, for example). A 16-bit MAC operation is two orders of magnitude more energy efficient than fetching its operands from the last level cache. Even worse are communications over off-chip channels. They require, approximately, a factor of 40x more energy than accesses to the last level cache, which is two to three orders of magnitude more than the energy dissipated by floating point operations and 16-bit MAC operations.

The takeaway message from such data is that the main source of power consumption in most CPUs is not calculations but rather transferring data and instructions over large distances on the chip, or even worse, off chip. Essentially, there is a significant tax on all data motion. From NVIDIA's perspective, GPUs lessen the sting of this tax through large amounts of data parallelism—that is, by amortizing the energy cost of instruction and data transfers (and their control) over many more data calculations.

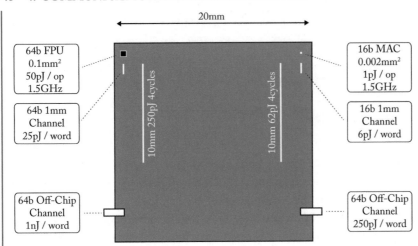

Figure 4.1: This figure depicts the energy cost of different computation and communication actions on a processor chip [40, 41, 104].

More broadly, this analysis points to the importance of considering instruction and data *motion*—not just logic and calculations—when building power-aware systems. The "first wave" of power optimizations focused on power optimizations within logic or arithmetic modules: turning off the multiplier when not in use, or clock-gating the upper bits of an adder for narrow-width operands. In contrast, the subsequent waves of power optimizations must focus on minimizing data motion and tailoring it to application needs, as subsequent sections discuss.

4.2   POWER AWARENESS IN ON-CHIP INTERCONNECT: TECHNIQUES AND TRENDS

With interprocessor interconnect dissipating more than 30% of on-chip power in many cases, its power efficiency has become a prominent focus of research attention.

4.2.1   BACKGROUND AND INDUSTRY STATE

Currently, most general-purpose CMPs still use cache-coherent hierarchies interconnected by buses to support cross-core communication on-chip. For example, Power7 uses eight 16-byte segmented buses [175]. Some designs are changing quickly as AMD and others push toward unified memory access between CPU and GPU [34, 118].

In general, as on-chip heterogeneity has increased, more diversity has emerged in on-chip connectivity. For example, Intel's Haswell (and its predecessors like Sandy Bridge [203]) has a ring network connecting the CPU cores, the GPU, and the last-level cache [72]. By clocking the CPU, GPU, and LLC domains independently, better power-performance tradeoffs are possible [173].

In essence, Intel has added more power gating and decoupled the ring-based interconnect and last-level cache from the CPU and GPU cores. For example, this potentially saves power by enabling the GPU to access the ring bus without waking the CPU cores. Expanding on this granularity, ultra-low-power Haswell mobile processors are equipped with deeper C-States (C8, C9, C10) that enable practically everything to be turned off when idle [88]. Thus, in a chip like Haswell, the segmented ring network itself is not highly novel, but its contribution to power savings and energy proportionality is quite significant.

Tilera [194] has popularized the mesh interconnect but offers relatively few concrete power results. They take the approach of multiple mesh networks to eliminate virtual channels and reduce buffering to a minimum. They contend that virtual channels are too expensive in power and area so multiple physical networks is more desirable.

Cavium 32-core Octeon II [107, 108] uses a crossbar interconnect arguing best trade-off of scalability and efficiency. Oracle has been sticking with an 8x9 crossbar in their T5 [148] but has not presented any details on implementation or power characteristics. To improve scaling, they implement a hierarchy, with a cluster of cores attached to one port of the crossbar. This saves power by avoiding the cost of routers and multi-hop traversals.

Intel has published power numbers and some discussion of power-efficient techniques for their prototypes, such as TeraFlops [79] and SCC [80, 81]. While many use TeraFlops mesh interconnect as an indication of the high contribution of interconnect power to overall chip power [79], the SCC uses a much smaller fraction of total chip power. SCC has many buffers and virtual channels (in contrast to Tilera's approach) and applies DVFS for power savings. Larrabee [166] used a ring interconnect with 512-bit links with very simple routers (no buffers in the routers) presumably for low-cost and energy efficiency. To the best of our knowledge Xeon Phi is using a similar interconnect structure as Larrabee.

4.2.2   POWER EFFICIENCY OF INTERCONNECT LINKS

One category of interconnect research looks to make existing networking approaches more power-efficient on a link-by-link basis. For example, researchers have explored both DVFS and on/off control for link power management. Some of the earliest work in this topic area [172], showed the large potential savings achievable through per-link DVFS approaches for networks on chip (NoC). Shang et al. envision multiple DVFS levels per link. When increasing link speed, the voltage is increased before the frequency, and when decreasing link speed, the transitions occur in the opposite order. This enables the link to function during the voltage transition, though not the frequency transition. The interesting challenge for this sort of link speed and power management is determining good policies by which the link speed can be decided. Their work uses a distributed history-based policy. Each router port predicts its future workload based on the recent past traffic demands. Of several possible indicators, the authors settle on link utilization and input buffer utilization as the preferred metrics. They report an average power savings of 4.6×, with some benchmarks seeing improvements up to 6.3×. In current technologies, individualized per-link voltage and clock domains may not be cost-effective, but coarser-grained zoned versions of such techniques may have promise. In addition to general solutions for homogeneous parallelism, one can also exploit speed scaling to optimize power on the interconnect between processors and accelerators or specialized functional units. Because one may be able to better predict required data rates and streaming in such usage scenarios, accurate speed balancing is quite possible.

Roughly at the same time, a series of papers by Soteriou and Peh [176] explored on/off links. Here, concerns about routability make for interesting design tradeoffs. When links are powered off, they are unusable, so either they must be turned back on quickly, or one must allow for alternative routing patterns to avoid them until they are powered up.

Another category of interconnect power optimizations comes from reducing the activity factor on the interconnect. For example, filtering cache requests to limit broadcasts can offer very large power savings. Research by Jerger et al. has shown how to build virtual multicast trees within on-chip mesh interconnects [97]. Subsequent work on Virtual Tree Coherence [51] shows how to use these multicast trees to implement cache coherence operations that are more efficient (from both power and performance perspectives) than prior work based on directories. One key to the success of this work is that it limits communication to just the sub-portions of the chip that need to be involved in each coherence operation. This saves power (and bandwidth) on links to uninterested cores. Compression has also been employed to reduce the amount of data that needs to be transferred over the interconnect. Jin et al. propose to use a table-based data compression technique that dynamically tracks value patterns in the traffic [98]. They are able to reduce the power consumption by 36% for a 16-core CMP while at the same time reducing the packet latency by 36%.

More recently, the field of power-aware interconnects has pushed toward full fabricated research prototypes [151, 155, 165, 170]. These works each demonstrate the orchestrated use of several power efficiency techniques in real hardware and report real-system chip-level power data.

4.2.3 EXPLOITING EMERGING TECHNOLOGIES TO IMPROVE POWER EFFICIENCY

Traditional "wired" on-chip interconnect approaches work well for local interconnect, but run into latency and power challenges for the increasingly long cross-chip wires. While improvements to these traditional approaches may help, additional, more forward-looking research on interconnect power seeks to augment traditional "wired" on-chip interconnect technologies or even entirely re-architect the interconnect by exploiting other options based on emerging technologies including three-dimensional (3D) integrated, on-chip photonics, or on-chip radio frequency (RF) communication. Unlike some prior topics, the "jury is still out" regarding which of these ideas will see widespread industry adoption and to what degree their power/performance characteristics will play a role in that decision. Carloni et al. [31] provide a good, brief introduction to the different techniques various pros and cons.

The topic of on-chip photonic networks has seen particular attention from architects [36, 113, 114, 171]. While optical communication itself is exceedingly power efficient, the electrical-optical translation is power hungry, as can be dynamic routing of the optical signals. Over the years, sequences of ideas have brought progress on these fronts. For example, Phastlane [36] is a low-latency optical crossbar network that uses predecoded source routing to transmit cache-line-sized packets several hops in a single clock cycle under contentionless conditions. If contention exists, the router reverts to using electrical buffers. The authors claim that Phastlane achieves 2× better network performance than a state-of-the-art electrical baseline while consuming 80% less network power. Pushing this progression even further, Kirman and Martínez proposed an *all-optical* scheme for wavelength-based routing in on-chip interconnect [114]. By using passive optical wavelength routers, significant area and power savings are possible compared to schemes relying on dynamic routers.

On-chip RF interconnect has seen less attention, but has appeal because of its potential for power savings with high performance. Chang et al. [33] propose a NoC architecture that is a hybrid of a traditional wireline mesh interconnect with an overlay of additional RF "wireless" connectivity. The RF interconnect is a high-bandwidth frequency-division multiple access (FDMA) transmission line that acts as a superhighway for fast data communication across longer chip distances. In addition to reduced latency, the authors show that NoC power dissipation using overlaid RF short-cuts is an order of magnitude less than in traditional wireline NoCs.

4.3 POWER AWARENESS IN DATA STORAGE: CACHES AND SCRATCHPADS

In addition to data motion and interconnects, data storage in caches and buffers represents a second aspect of data's energy "tax" on computation. We cover key techniques for improving energy efficiency in this section.

At an intuitive level, local storage like caches and scratchpads seem in many ways to offer inherent power efficiency. By serving the most commonly referenced items from a small block of storage close to the computation, level-one (L1) caches or scratchpads can help filter requests such that fewer of them need to expend the extra energy required to go out to larger, more distant, and higher energy-per-access levels of the memory hierarchy. The memory hierarchy can still account for a substantial part of the total chip power as illustrated by the pie chart in Figure 4.2. The power consumption of the three levels of caches is greater than the power consumption of the eight cores on the chip. Clearly, it is important to implement techniques that reduce the average memory access energy. We discuss such techniques in the subsections that follow.

4.3.1 CACHE HIERARCHIES AND POWER EFFICIENCY

While L1 instruction and data caches seem to offer great opportunities for "filtering" references to reduce higher-energy accesses to larger downstream caches, they often have large energy-per-

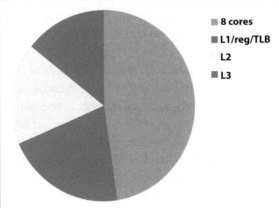

8 cores

L1/reg/TLB

L2

L3

Figure 4.2: Power breakdown of an eight core server chip [77].

access. This is because L1 caches are highly optimized for performance, so their transistors are sized large for fast access times. In addition, typical associative L1 cache designs often activate all ways in parallel to find data or instructions as quickly as possible. In contrast, later levels of the cache hierarchy may be optimized for capacity rather than access speed, so their energy-per-access need not grow commensurate with the increased cache capacity.

Beyond this simple speed tradeoff, since software is typically less latency-sensitive to level-two (L2) and level-three (L3) cache designs, they offer better opportunities for infusing power efficiency into the design. For example, accesses to L2 and L3 caches are commonly "phased" such that the tags are read and compared before accessing the data. Knowing which way the data resides in makes it possible to enable only one of the data ways when accessing the data memories. Note that the data memories are much larger and more power hungry than the tag memories, as they store more bits (a whole cache line, which is often 256 or more bits, compared to a tag, which is fewer than 64 bits).

Accesses to the L1 cache can be reduced by introducing a so-called filter or L0 cache [111, 112]. An L0 cache is smaller than the L1 cache and hence more power efficient. Due to its small size L0 caches have a relatively high miss rate, which causes a performance penalty as an additional cycle is required to access the L1 cache. Bardizbanyan et al. proposed a technique to speculatively access an L0 data cache in parallel with its address calculation to not only eliminate the performance penalty but also provide a small performance improvement compared to not having an L0 data cache [15]. L0 caches have been adopted by Qualcomm, which use a direct mapped L0 cache in their Krait architecture [115].

Loop caches are similar to filter caches in that a small and thus power-efficient structure is used to reduce accesses to the L1 instruction cache [120]. As the name implies, loop caches exploit the repetitive pattern of loops where the same instructions are repeatedly fetched multiple times. When a backward direct transfer of control is encountered the loop cache starts to store instructions in the hope that the same transfer of control will be encountered again. If the transfer happens again the stored instructions are fetched from the loop cache instead of accessing the L1 cache. Instruction fetches from the loop cache also have the added advantage of disabling accesses to the instruction translation lookaside buffer (ITLB). The tagless hit instruction cache (TH-IC) also exploits the repetitive behavior of instruction accesses to store the most commonly encountered instructions in a small structure [61, 74]. By maintaining metadata related to the instructions guarantees can be provided regarding if an instruction resides in the TH-IC. When an instruction is guaranteed to reside in the TH-IC accesses to the instruction cache and other fetch related structures (e.g., the ITLB) can be avoided.

Caches fetch complete lines of data to take advantage of spatial locality. This causes more data than what is accessed by an application to be fetched and stored in the caches. Pujar and Aggarwal showed that the cache utilization (the fraction of useful data over all fetched data) is as low as 57% on average for the SPEC 2K benchmarks [157]. The low utilization causes unnecessary data movement across on-chip and off-chip interconnects as well as costly reads and writes to cache memories.

4.3.2   CACHE ASSOCIATIVITY AND ITS IMPLICATION ON POWER

L2 and L3 caches commonly have a much higher associativity than the L1 cache to improve their hit rate. This can result in high access costs even in the case when phased accesses are used as many parallel tag lookups are required. Several techniques have been proposed to reduce the number of tag comparisons that are performed.

For example, way prediction predicts which way of a set associative cache is most likely to contain the requested data, and starts the access by looking only in that way for the data. This saves expending power on the other cache ways when prediction is accurate [84, 156]. When the way is mispredicted a performance penalty is incurred, as the remaining ways need to be accessed in the following cycle, which reduces the energy benefits. Also, way prediction needs to be performed before the actual access to the memories, which results in either a longer access time where the prediction and memory access are performed in sequence or specialized memory implementations are required that can disable the initiated memory access once the prediction is completed. Multicore systems and multiprogram workloads reduce the locality in the memory accesses seen by shared caches in a system as accesses from independent program executions are interleaved. This makes it challenging to accurately predict the way of an access for shared caches. The higher locality at the L1 cache makes predictions easier to perform, but the often stringent timing constraints for L1 cache accesses might make it difficult to incorporate such techniques. A private L2 cache has high locality, often more ways than the L1 cache, and relaxed timing requirements, making it a prime candidate for techniques to predict which way to access first.

Way halting is another technique for reducing the number of tag comparisons [206]. In way halting partial tags are stored in a fully associative memory (the halt tag array) with as many ways as there are sets in the cache. The halt tag array is tightly coupled with the word line decoder of the memories storing the tag and data. In parallel with decoding the word line address the partial tag is searched in the halt tag array. Only for the set where a partial tag match is detected can the word line be enabled by the word line decoder. This halts access to ways that cannot contain the data as determined by the partial tag comparison. There is a tradeoff between how many bits to use for the partial tag. The more bits that are stored the more likely it is that the partial tag will not match and the way will be halted, but the access to the halt tag array will become more costly in terms of energy and could increase the access delay. Way halting requires specialized memory implementations that might have a negative impact on the maximum operational frequency.

Bardizbanyan et al. propose a technique that speculatively accesses the tags a cycle earlier than for a conventional L1 data cache by using the index of the base address register [16]. This type of speculative access is only performed when the displacement for the address calculation is smaller than the cache line offset. With a small displacement it is likely that the index will not change during the address calculation. If the index does not change during the address calculation, i.e., the speculation is successful, then energy is saved by only accessing a single way of data. If the index does change then the tags is accessed a second time in parallel with the data. A failed speculation causes a small energy overhead but no performance penalty. Bardizbanyan et al. also propose to dynamically track load-use dependencies in a small structure to enable early load data dependence detection (ELD^3) [17]. If a load has been encountered already then the ELD^3 metadata contains information on if a use of the loaded value immediately follows the load. If that is the case then both the tags and data are accessed in parallel, otherwise the access is phased such that only a single data way is read to save energy.

Not all applications (or phases of an application) benefit from a highly associative cache. For such applications the associativity causes high power consumption without providing significant performance gains. Dropsho et al. [47] propose to use a dynamically reconfigurable cache where the ways in a cache are divided into two groups. On an access the ways in the first group are accessed and if a miss is encountered the second group of ways is accessed. Cache lines are migrated between the two groups such that the most frequently accessed cache lines reside in the first group of ways. The number of ways in the first and second group can be dynamically configured and a scheme based on least recently used (LRU) information is used to decide how many ways to allocate to the first and second group depending on application behavior. Way concatenation [205] is another method of dynamically adapting the associativity of the cache. Pairs of ways are concatenated into a single way when reducing the associativity of the cache. The combined way has twice the number of sets as the two individual ways, thus the cache capacity is not changed. Pairs of ways can be further concatenated until a direct mapped cache with a single way remains.

4.3.3   CACHE RESIZING AND STATIC POWER

Intel resizes the last-level cache from an 8 MB 16-way cache to 512 kB two-way cache in their IvyBridge architecture [95]. When low activity workloads are detected 14 of the ways are flushed

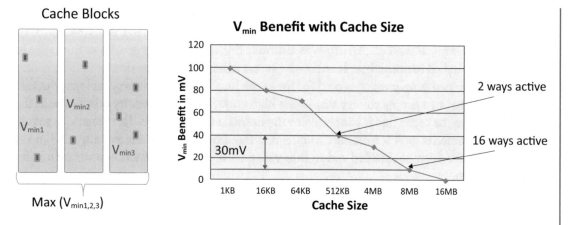

Figure 4.3: V_{min} is determined by the cell with the highest V_{min} across all active memories. Reduction in V_{min} when reducing the number of ways from 16 to two in Intel IvyBridge [95].

and put into a sleep mode that reduces their static power consumption. The remaining two cache ways can be operated at a lower minimum voltage, which reduces the dynamic power even further. The minimum voltage that a cache can be operated at is dictated by defective (bad) memory cells that cannot operate correctly at a voltage lower than a certain minimum voltage (V_{min}). Such defective cells are distributed across the memory arrays of the cache and the cell requiring the highest voltage dictates the minimum operational voltage for the whole cache. When the cache size is reduced there are fewer defects and the minimum voltage can be reduced (see Fig. 4.3) [95].

Wilkerson et al. propose a technique that trades cache capacity against the minimum voltage a cache can be operated at [195, 196]. During normal high-voltage operation the cache can be used at its full capacity while the capacity is reduced during low-voltage operation. With each tag a mask is stored that tracks defective memory cells (or words) that does not operate correctly at the lower voltage. A pair of cache lines are used in low-voltage mode and the mask indicates which cache line holds the correct data. This technique masks defective memory cells, which enables the use of a lower minimum voltage.

Caches are large and have relatively low activity compared to logic. This is especially true for the last-level cache (LLC) that can store up to tens of megabytes with only a small fraction being accessed at a single point of time. With shrinking transistor sizes static power is an ever greater concern and constitutes a substantial fraction of total cache power. In so-called drowsy caches static power is reduced by putting cache lines into a low-power (drowsy) mode [58]. In this drowsy mode a cache line is powered at a level that the line retains its data but it is not possible to access it. All cache lines are periodically put in the drowsy mode and only when a cache line is accessed it is switched to normal mode and kept in this mode for the remaining time of the period. This incurs a minor delay penalty while keeping the majority of the cache lines in a drowsy state.

A read access to a cache line in drowsy mode would corrupt the data. To avoid data corruption a drowsy bit is stored for each cache line and if the bit is set then the word line is disabled such that the cache line is not accessible. An access to a drowsy line incurs a delay penalty as the cache line needs to be switched to normal mode and a new access to be initiated. The drowsy bit also controls switches to the two supply voltages used for the cache line. When the drowsy bit is set the cache line is connected to a supply voltage that is just high enough for the memory cells to retain their data, otherwise the cache line is connected to the normal supply voltage. This technique requires specialized memories as it is necessary to integrate the drowsy bit into the memory design such that the word line can be disabled and to integrate dual supply voltages with power switches for each cache line.

4.3.4   CACHE COHERENCE

Shared memory is the predominant method of communication in contemporary multi-core systems and is supported by all major processor vendors. Communication through shared memory is performed through load and store operations to a shared address space. Cache coherence protocols assure that private caches are kept coherent and reflect the global state of the shared memory. Cache coherence can be managed in software, but mainly due to ease of programmability most systems implement coherence in hardware. Here we focus on hardware coherence protocols and techniques to improve their energy efficiency.

Snoop-Based Protocols
In snoop-based protocols a cache miss in the last-level private cache (e.g., in a private L2 cache) results in a data request being broadcasted to all other cores in the system. At each core a lookup is performed in their private caches and if the data are found then the data are transferred back to the core initiating the data request. Broadcasting coherence requests can result in significant on-chip network traffic and multiple parallel lookups in the cores private caches. To reduce bandwidth requirements and power, several techniques to limit the number of recipients of a broadcast have been developed. This can be done either at the receiver's end by filtering out requests to a core's private cache(s) or at the core, initiating the request by filtering out if a request should be sent and to which cores.

Moshovos et al. use what they call a Jetty to filter requests to the private cache(s) of a core [141]. A coherence request first checks in the Jetty to determine if the data might reside in the private cache(s). The Jetty can either be an exclude-Jetty and store information about recent coherence requests that missed in the private cache(s) and if the same request is seen again then no lookups are performed, or be an include-Jetty and store information of all the blocks that exist in the private cache(s). Only if the Jetty indicates that the information exists in the private cache(s) a lookup is performed. Both the exclude and include Jetty gives a prediction regarding if the data resides in the cache or not. Both types can be combined to increase the prediction success rate. In many cases the requested data does not reside in the private cache(s) of a core, and power hungry lookups can be avoided. A coherence request is still broadcasted to all the Jettys (one per core) and latency increases as the Jetty is accessed before the private cache(s) for the case when the Jetty indicates that the data resides in the private caches.

Ekman et al. propose to reduce snoop traffic by extending the TLB with an auxiliary structure containing a page-sharing table [50]. The page-sharing table keeps track of pages that are shared by the core. For each shared page a sharing vector is kept that indicates which of the other cores also share the page. The page-sharing vector is broadcasted with the coherence request, and only those cores that are marked in the sharing vector need to access their private cache(s). The sharing vectors are kept up to date by querying all the tables when a core loads a new entry into its table. This technique reduces cache accesses at the expense of tracking global state of shared pages. A simple extension to also reduce interconnect bandwidth would be to not propagate coherence requests over the on-chip interconnect to cores that do not share the page as determined by the sharing vector.

RegionScout keeps track at the core level of continuous, aligned memory regions that are a power of two in size and have no shared data [140]. When a core makes a coherence request it receives region-level sharing information in addition to the requested data. If a region is identified as not shared then subsequent requests from the same core for any data within that region will not cause any coherence requests. A region is not shared if a coherence request results in no other core returning the data. A coherence request from a different core to a region will mark that region as being shared. RegionScout can reduce both network traffic and cache lookups. Accesses to regions that are marked as shared results in coherence requests being broadcasted to all the cores.

The ARM CoreLink CCN-504 cache coherent network [9] has support for up to 16 cores divided into four clusters with four cores in each cluster. The network has a snoop-based cache coherence protocol with snoop filter functions that removes the need to broadcast coherence messages. The Blue Gene/P supercomputer from IBM implements a snoop filter that combines a snoop cache, a stream register filter [162], and a range filter to reduce the number of snoop requests by up to 99% [161]. The snoop cache keeps track of what is not in the cache while the stream register filter keeps track of what is in the cache. The range filter unconditionally filters all coherence requests within a specified address range. The snoop cache is a version of the exclude-Jetty (described above). To increase the snoop cache's efficiency each entry stores a presence vector that encodes information about 32 consecutive, aligned cache lines of the L1 data cache.

Directory-Based Protocols

Directory-based protocols can reduce network traffic and be more scalable than snoop-based protocols. However, the directories incur area overheads and contribute both to static and dynamic power. Several techniques have been proposed to either make the directories themselves more efficient or to completely avoid accessing them when not necessary.

The tagless coherence directory [204] use a grid of Bloom filters instead of tags to perform lookups in the directory. The Bloom filters are smaller than the tags, which results in reduced static and dynamic power. In proximity coherence directory accesses are avoided by snooping the caches of neighboring cores [20]. On a so-called proximity hit the data are found in at least one of the neighboring core's L1 cache and are in a suitable state (e.g., the data are in shared mode on a load). In this case the request can be immediately serviced without accessing the directory, and network traffic is kept local. Forwarding data between neighboring cores without involving the directory causes challenges for the design of the directory as it is not aware of new cores sharing the data. Therefore, when an L1 cache line that has been shared with other cores is evicted an update has to be sent to the directory with all the cores that the line has been shared with due to proximity requests. When a core receives an invalidate message it needs to forward the message to all the cores that the cache line has been forwarded to. If a core requires exclusive access it needs to invalidate the cache line in those cores it has forwarded it to and send an upgrade message to the directory. The evaluation of proximity coherence is done with an assumption of a shared L2 architecture. For architectures with deeper private cache hierarchies either a proximity request could be performed in sequence at the different private levels, which would increase the proximity hit rate but increase the latency on misses, or only snoop the L1 before accessing the directory. TurboTag filters unnecessary directory accesses by using a compact structure based on a counting Bloom filter that keeps track of block addresses that are used in the directory [128]. The filter is queried first and only if it indicates that the block potentially exists in the directory; then an access is made to the conventional directory that contains sharer information. If the filter indicates that the information does not exist in the directory then the access is skipped, which reduces access power as the filter is a much smaller structure than the complete directory.

4.3.5 THE POWER IMPLICATIONS OF SCRATCHPAD MEMORIES

With GPUs and specialized accelerators seeing renewed attention from computer architects, the prominence of programmer-managed memories called scratchpads has also risen. These small local memories have the potential for improving the power efficiency of data motion by engaging programmer or compiler help in orchestrating the selection of data to be placed into them. As a result, wasted energy due to repeated transfers (cache thrashing) or due to cache line granularity effects can be reduced. Another advantage of scratchpads relative to caches is that they do not require any tags. Scratchpads are therefore inherently smaller and have less static power for equivalent amount of data storage. Scratchpads also have less dynamic power as no tags have to be read and compared and only a single data entry is read while multiple ways are read in parallel for associative caches.

In some cases, the work has been focused on compiler or programmer optimizations for selecting data to be placed in the scratchpad [21, 125, 126, 180, 187, 188]. In other cases, more holistic studies have considered optimizations for unifying and coordinating the use of cache, scratchpad, and register file storage. In particular, Gebhart et al. propose to unify the register file, cache, and scratchpad of a GPU core into a shared resource [63]. The unified local memory consists of a set of memory banks and a crossbar interconnect, which enables resources to be dynamically allocated as registers, scratch area, or cache size depending on kernel requirements. A compile time managed multi-level register file is used to reduce contention at the shared resource and to improve access times to registers [62]. Dynamic resizing is supported between co-operative thread arrays (CTAs) as registers and shared memory (scratchpad) are not persistent across CTAs. The cache is write through and therefore does not contain any dirty data to evict if the cache would be reduced in size.

4.4   CHAPTER SUMMARY

The complexity of computer systems is increasing. The number of cores are slowly rising and are often coupled with accelerators like GPUs and cryptographic engines. The general trend is also to integrate more functionality onto the same chip to reduce total system cost but also to improve performance and reduce energy usage. One such example is when AMD started integrating the memory controller and the north bridge with their processors, which made AMD's Hammer architecture [105] superior in performance until Intel followed suit. Today the integration has gone further with functionality like USB and Ethernet being integrated into what has been termed "system on chips" (SoCs). This is especially true for the embedded domain.

The increasing number of CPU, GPU, and IP (intellectual property, e.g., USB) cores puts high demands on the coordination and sharing of data. The on-chip interconnect and memory hierarchy is therefore playing an ever greater role in the performance and energy usage of a system. The focus has therefore started to shift from energy-efficient computation to energy-efficient data movement and we are likely to see a trend that continues to emphasize the importance of data movement and storage.

CHAPTER 5

Conclusions

Over the past two decades, perspectives on power efficiency have shifted tremendously within the field of computer architecture. In the early 1990s—the end of the bipolar/ECL era for microprocessors—power dissipation had become something of a concern in industry chip designs as wattages surpassed the 100W mark [99], but did not get much attention in research or early-stage development. The sense at the time was that power concerns were mainly to be addressed by device and circuit levels instead, leaving architects to focus on performance. The late 1990s represented the onset of a broad awareness of computer architecture's role to play in power-aware design, and the years that followed saw a burst of research and product ideas on this topic—many of which were highlighted by S. Kaxiras and M. Martonosi [103].

Over the past ten years, from roughly 2005 to the time of this writing, power-aware architecture has seen another shift. In particular, the attention to module-by-module power efficiency ideas has shifted toward more holistic power/performance tradeoffs that permeate entire chip designs. For example, the trends of parallelism, heterogeneity, and specialization are all fundamentally driven by power concerns, and by power's influence on system cost and reliability. The goal of this book has been to convey the drivers for these current trends, as well as the research ideas and design techniques developed in response to them.

5.1 FUTURE TRENDS: TECHNOLOGY CHALLENGES AND DRIVERS

The architecture layer has always acted as a crucial mediator between the technology opportunities and challenges "from below" and the software and application needs "from above." For years, architects have occupied the enviable position of leveraging and amplifying the technology bonanza afforded to us by Moore's Law and Dennard scaling trends. Successive semiconductor generations will offer some improvements in transistor fabrication techniques in order to mitigate the emerging challenges of leakage energy and process variations. Beyond these measures, there is at best speculation concerning what follows next: device researchers explore new technology options, but no clear prospects have emerged.

With the slowing or stopping of beneficial scaling trends for transistor counts and voltage levels, architecture has dramatically shifted gears and will continue to do so. The response to Dark Silicon trends represents one avenue of ongoing and future work. Another is an increased focus on 3D and other techniques to bring data closer to memory. Looking forward, techniques for co-locating processing near storage [43] represent promising techniques for reducing data motion energy by preprocessing data while it is still located at storage, before it incurs power to move it toward other compute nodes. For emerging data-intensive workloads like search and analytics, sifting through the data near "where it lives" is particularly well-suited. Such techniques offer considerable advantages in power efficiency, and likely represent a major future thrust for power-aware systems research.

5.2   FUTURE TRENDS: EMERGING APPLICATIONS AND DOMAINS

Architectures in coming years will also need to respond to fast-moving changes in software applications and usage domains. Already, mobile computing represents a dominant mode for computer usage, with the number of smartphones on earth fast approaching the number of people [60]. As the numbers of deployed mobile devices have skyrocketed, they now are in a position to drive computer architecture trends simply due to their sheer sales volume. The most mature solutions for exploiting heterogeneous computing have been pioneered in the mobile and embedded space, with influences now percolating into server-class systems.

Moving forward, the next major burst of design and sales volume is expected to be in the deeply embedded or "Internet of Things" space [131]. The emergence of this space will have sweeping implications on computer architecture and systems research in general, and on power-aware computing in particular.

Many such devices operate with tightly constrained power or thermal envelopes that are fundamentally imposed by the application scenario itself, rather than by cost or other secondary concerns. For example, medically implanted sensors must abide by strict power and thermal limits to avoid injuring the tissue near where they are embedded [189]. Likewise, devices that operate on energy-harvested budgets must operate using very little energy and must be designed to operate intermittently and reactively in the face of highly variable energy supplies [149].

5.3   FINAL SUMMARY

Given power's prominent place in computer architecture research as a first-class design constraint, it has become increasingly impossible to offer a truly comprehensive synthesis of power-aware computing research—because now all architecture research has become power-aware. Instead, this book serves to offer highlights of promising and timely techniques and trends. Our emphasis on those developed in the past five years is a judicious effort to offer a timely update on S. Kaxiras and M. Martonosi's book [103]. Nonetheless, our field is fast-moving and power's prominence is only growing; we look forward to watching as new power tradeoffs emerge and new techniques are proposed to address them.

Bibliography

[1] Y. Abe, H. Sasaki, M. Peres, K. Inoue, K. Murakami, and S. Kato, "Power and performance analysis of GPU-accelerated systems," in *Proceedings of the USENIX Conference on Power-Aware Computing and Systems*, Oct. 2012, pp. 10–10. 40

[2] C. Akass, "ARM wrestles with dark silicon," Mar. 2010, web page. [Online]. Available: http://www.theinquirer.net/inquirer/feature/1598659/arm-wrestles-dark-silicon 7

[3] AMD, "AMD Turbo Core technology," web page. [Online]. Available: http://www.amd.com/en-gb/innovations/software-technologies/turbo-core 17, 34

[4] AMD, "AMD PowerNow!™ technology dynamically manages power and performance," Nov. 2000, AMD white paper. [Online]. Available: http://www.amd-k6.com/wp-content/uploads/2012/07/24404a.pdf 22

[5] AMD, "AMD Fusion APU era begins," Jan. 2011, web page. [Online]. Available: http://www.amd.com/en-us/press-releases/Pages/amd-fusion-apu-era-2011jan04.aspx 39

[6] AMD, "Compute cores," Jan. 2014, AMD white paper. [Online]. Available: http://www.amd.com/Documents/Compute_Cores_Whitepaper.pdf 39

[7] G. M. Amdahl, "Validity of the single processor approach to achieving large scale computing capabilities," in *Proceeding of the Spring Joint Computer Conference*, Apr. 1967, pp. 483–485. DOI: 10.1145/1465482.1465560. 5

[8] M. Annavaram, E. Grochowski, and J. Shen, "Mitigating Amdahl's law through EPI throttling," in *Proceedings of the International Symposium on Computer Architecture*, Jun. 2005, pp. 298–309. DOI: 10.1145/1080695.1069995. 25

[9] ARM, "CoreLink CCN-504 cache coherent network," web page. [Online]. Available: http://www.arm.com/products/system-ip/interconnect/corelink-ccn-504-cache-coherent-network.php 55

[10] ARM, "Cortex™-A15 MPCore™," ARM technical reference manual. [Online]. Available: http://infocenter.arm.com/help/index.jsp?topic=/com.arm.doc.ddi0438h/index.html 34

[11] ARM, "Cortex™-A7 MPCore™," ARM technical reference manual. [Online]. Available: http://infocenter.arm.com/help/index.jsp?topic=/com.arm.doc.ddi0464d/I1007542.html 34

[12] ARM, "Heterogeneous multi-processing solution of Exynos 5 Octa with ARM® big.LITTLE™technology," 2013, white paper. [Online]. Available: http://www.arm.com/files/pdf/Heterogeneous_Multi_Processing_Solution_of_Exynos_5_Octa_with_ARM_bigLITTLE_Technology.pdf 7

[13] M. Arora, S. Nath, S. Mazumdar, S. Baden, and D. Tullsen, "Redefining the role of the CPU in the era of CPU-GPU integration," *IEEE Micro*, vol. 32, no. 6, pp. 4–16, 2012. DOI: 10.1109/MM.2012.57. 40

[14] T. Austin, V. Bertacco, D. Blaauw, and T. Mudge, "Opportunities and challenges for better than worst-case design," in *Proceedings of the Asia and South Pacific Design Automation Conference*, Jan. 2005, pp. 2–7. DOI: 10.1145/1120725.1120878. 42

[15] A. Bardizbanyan, M. Själander, D. Whalley, and P. Larsson-Edefors, "Designing a practical data filter cache to improve both energy efficiency and performance," *ACM Transactions on Architecture and Code Optimization*, vol. 10, no. 4, pp. 54:1–54:25, Dec. 2013. DOI: 10.1145/2541228.2555310. 50

[16] A. Bardizbanyan, M. Själander, D. Whalley, and P. Larsson-Edefors, "Speculative tag access for reduced energy dissipation in set-associative L1 data caches," in *Proceedings of the IEEE International Conference Computer Design*, Oct. 2013, pp. 302–308. DOI: 10.1109/ICCD.2013.6657057. 52

[17] A. Bardizbanyan, M. Själander, D. Whalley, and P. Larsson-Edefors, "Reducing set-associative L1 data cache energy by early load data dependence detection ELD3," in *Proceedings of the Conference on Design, Automation and Test in Europe*, Mar. 2014, pp. 82:1–82:4. DOI: 10.7873/DATE.2014.095. 52

[18] L. A. Barroso and U. Hölzle, "The case for energy-proportional computing," *Computer*, vol. 40, no. 12, pp. 33–37, Dec. 2007. DOI: 10.1109/MC.2007.443. 1, 37

[19] L. A. Barroso and U. Hölzle, "The datacenter as a computer: An introduction to the design of warehouse-scale machines," in *Synthesis Lectures on Computer Architecture*. Morgan & Claypool Publishers, Jul. 2009. DOI: 10.2200/S00516ED2V01Y201306CAC024 2

[20] N. Barrow-Williams, C. Fensch, and S. Moore, "Proximity coherence for chip multiprocessors," in *Proceedings of the International Conference on Parallel Architectural and Compilation Techniques*, Sep. 2010, pp. 123–134. DOI: 10.1145/1854273.1854293. 56

[21] M. M. Baskaran, U. Bondhugula, S. Krishnamoorthy, J. Ramanujam, A. Rountev, and P. Sadayappan, "Automatic data movement and computation mapping for multi-level parallel architectures with explicitly managed memories," in *Proceedings of the ACM SIGPLAN Symposium on Principles and Practice of Parallel Programming*, Feb. 2008, pp. 1–10. DOI: 10.1145/1345206.1345210. 56

[22] A. Bhattacharjee and M. Martonosi, "Thread criticality predictors for dynamic performance, power, and resource management in chip multiprocessors," in *ACM SIGARCH Computer Architecture News*, vol. 37, no. 3, Jun. 2009, pp. 290–301. DOI: 10.1145/1555815.1555792. 28

[23] C. Bienia, S. Kumar, J. P. Singh, and K. Li, "The PARSEC benchmark suite: Characterization and architectural implications," in *Proceedings of the International Conference on Parallel Architectural and Compilation Techniques*, Oct. 2008, pp. 72–81. DOI: 10.1145/1454115.1454128. 32

[24] M. Bohr and K. Mistry, "Intel's revolutionary 22 nm transistor technology," May 2011, Intel presentation. [Online]. Available: http://download.intel.com/newsroom/kits/22nm/pdfs/22nm-details_presentation.pdf 3, 14

[25] S. Borkar, "Design challenges of technology scaling," *IEEE Micro*, vol. 19, no. 4, pp. 23–29, Jul. 1999. DOI: 10.1109/40.782564. 1

[26] S. Borkar and A. A. Chien, "The future of microprocessors," *Communications of the ACM*, vol. 54, no. 5, pp. 67–77, May 2011. DOI: 10.1145/1941487.1941507. 3, 4, 5, 6, 12

[27] D. Brooks, V. Tiwari, and M. Martonosi, "Wattch: A framework for architectural-level power analysis and optimizations," in *Proceedings of the International Symposium on Computer Architecture*, Jun. 2000, pp. 83–94. DOI: 10.1145/342001.339657. 21

[28] N. Brookwood, "AMD Fusion™ family of APUs: Enabling a superior, immersive PC experience," Mar. 2010, AMD white paper. [Online]. Available: http://sites.amd.com/kr/Documents/48423B_fusion_whitepaper_WEB.pdf 7

[29] D. Burger, S. W. Keckler, K. S. McKinley, M. Dahlin, L. K. John, C. Lin, C. R. Moore, J. Burrill, R. G. McDonald, W. Yoder, and t. T. Team, "Scaling to the end of silicon with EDGE architectures," *Computer*, vol. 37, no. 7, pp. 44–55, Jul. 2004. DOI: 10.1109/MC.2004.65. 37

[30] Q. Cai, J. González, R. Rakvic, G. Magklis, P. Chaparro, and A. González, "Meeting points: Using thread criticality to adapt multicore hardware to parallel regions," in *Proceedings of the International Conference on Parallel Architectural and Compilation Techniques*, Oct. 2008, pp. 240–249. DOI: 10.1145/1454115.1454149. 28

[31] L. P. Carloni, P. Pande, and Y. Xie, "Networks-on-chip in emerging interconnect paradigms: Advantages and challenges," in *Proceedings of the ACM/IEEE International Symposium on Networks-on-Chip*, May 2009, pp. 93–102. DOI: 10.1109/NOCS.2009.5071456. 48

[32] A. Chandrakasan, S. Sheng, and R. Brodersen, "Low-power CMOS digital design," *IEEE Journal of Solid-State Circuits*, vol. 27, no. 4, pp. 473–484, 1992. DOI: 10.1109/4.126534. 11

[33] M. F. Chang, J. Cong, A. Kaplan, M. Naik, G. Reinman, E. Socher, and S.-W. Tam, "CMP network-on-chip overlaid with multi-band RF-interconnect," in *Proceedings of the International Symposium High-Performance Computer Architecture*, 2008, pp. 191–202. DOI: 10.1109/HPCA.2008.4658639. 49

[34] H. Chu, "AMD heterogeneous uniform memory access," 2013, AMD presentation. [Online]. Available: http://events.csdn.net/AMD/GPUSat%20-%20hUMA_june-public .pdf 46

[35] E. S. Chung, J. D. Davis, and J. Lee, "LINQits: Big data on little clients," in *Proceedings of the International Symposium on Computer Architecture*, Jun. 2013, pp. 261–272. DOI: 10.1145/2485922.2485945. 41

[36] M. J. Cianchetti, J. C. Kerekes, and D. H. Albonesi, "Phastlane: a rapid transit optical routing network," in *Proceedings of the International Symposium on Computer Architecture*, 2009, pp. 441–450. DOI: 10.1145/1555815.1555809. 49

[37] Committee on Sustaining Growth in Computing Performance; National Research Council, *The Future of Computing Performance: Game Over or Next Level?*, S. H. Fuller and L. I. Millett, Eds. The National Academies Press, 2011, pp. 55. [Online]. Available: http://www.nap.edu/catalog.php?record_id=12980 4

[38] J. Cong, M. A. Ghodrat, M. Gill, B. Grigorian, and G. Reinman, "CHARM: a composable heterogeneous accelerator-rich microprocessor," in *Proceedings of the International Symposium on Low Power Electronics and Design*, 2012, pp. 379–384. DOI: 10.1145/2333660.23337479. 40

[39] G. Contreras, M. Martonosi, J. Peng, G.-Y. Lueh, and R. Ju, "The XTREM power and performance simulator for the Intel XScale core: Design and experiences," *ACM Transactions on Embedded Computing Systems*, vol. 6, no. 1, pp. 1–25, Feb. 2007. DOI: 10.1145/1210268.1210272. 21, 22

[40] W. Dally, "The future of GPU computing," 2009, keynote at SC. [Online]. Available: http://www.nvidia.com/content/GTC/documents/SC09_Dally.pdf 45, 46

[41] W. J. Dally, "GPU computing to exascale and beyond," 2010, keynote at SC. [Online]. Available: http://www.nvidia.com/content/PDF/sc_2010/theater/Dally_SC10.pdf 39, 46

[42] R. Dennard, F. Gaensslen, V. Rideout, E. Bassous, and A. LeBlanc, "Design of ion-implanted MOSFET's with very small physical dimensions," *IEEE Journal of Solid-State Circuits*, vol. 9, no. 5, pp. 256–268, 1974. DOI: 10.1109/JSSC.1974.1050511. 1, 2

[43] P. Dlugosch, D. Brown, P. Glendenning, M. Leventhal, and H. Noyes, "An efficient and scalable semiconductor architecture for parallel automata processing," *IEEE Transactions on Parallel and Distributed Systems*, vol. 25, no. 12, pp. 3088–3098, Dec. 2014. DOI: 10.1109/TPDS.2014.8. 59

[44] J. Donald and M. Martonosi, "Techniques for multicore thermal management: Classification and new exploration," in *Proceedings of the International Symposium on Computer Architecture*, 2006, pp. 78–88. DOI: 10.1145/1150019.1136493. 16, 25

[45] M. Dong and L. Zhong, "Self-constructive high-rate system energy modeling for battery-powered mobile systems," in *Proceedings of the International Conference on Mobile Systems, Applications, and Services*, 2011, pp. 335–348. DOI: 10.1145/1999995.2000027. 1

[46] J. Donovan, "ARM CTO: Warns of 'dark silicon'," Mar. 2010, web page. [Online]. Available: http://www.eetimes.com/author.asp?section_id=36&doc_id=1265557 32

[47] S. Dropsho, A. Buyuktosunoglu, R. Balasubramonian, D. Albonesi, S. Dwarkadas, G. Semeraro, G. Magklis, and M. Scottt, "Integrating adaptive on-chip storage structures for reduced dynamic power," in *Proceedings of the International Conference on Parallel Architectural and Compilation Techniques*, Sep. 2002, pp. 141–152. DOI: 10.1109/PACT.2002.1106013. 52

[48] K. Du Bois, S. Eyerman, J. B. Sartor, and L. Eeckhout, "Criticality stacks: Identifying critical threads in parallel programs using synchronization behavior," in *Proceedings of the International Symposium on Computer Architecture*, 2013, pp. 511–522. DOI: 10.1145/2485922.2485966. 29

[49] P.-E. Duhamel, J. Porter, B. Finio, G. Barrows, D. Brooks, G.-Y. Wei, and R. Wood, "Hardware in the loop for optical flow sensing in a robotic bee," in *Proceedings of the IEEE/RSJ International Conference Intelligent Robots and Systems*, 2011, pp. 1099–1106. DOI: 10.1109/IROS.2011.6095160. 40

[50] M. Ekman, P. Stenström, and F. Dahlgren, "TLB and snoop energy-reduction using virtual caches in low-power chip-multiprocessors," in *Proceedings of the International Symposium on Low Power Electronics and Design*, Aug. 2002, pp. 243–246. DOI: 10.1109/LPE.2002.146746. 55

[51] N. D. Enright Jerger, L.-S. Peh, and M. H. Lipasti, "Virtual tree coherence: Leveraging regions and in-network multicast trees for scalable cache coherence," in *Proceedings of the IEEE/ACM Annual International Symposium on Microarchitecture*, 2008, pp. 35–46. DOI: 10.1109/MICRO.2008.4771777. 48

[52] D. Ernst, N. S. Kim, S. Das, S. Pant, R. Rao, T. Pham, C. Ziesler, D. Blaauw, T. Austin, K. Flautner, and T. Mudge, "Razor: A low-power pipeline based on circuit-level timing

speculation," in *Proceedings of the IEEE/ACM Annual International Symposium on Microarchitecture*, 2003, pp. 7–18. 8, 42

[53] H. Esmaeilzadeh, A. Sampson, L. Ceze, and D. Burger, "Neural acceleration for general-purpose approximate programs," in *Proceedings of the IEEE/ACM Annual International Symposium on Microarchitecture*, Dec. 2012, pp. 449–460. DOI: 10.1109/MICRO.2012.48. 42

[54] H. Esmaeilzadeh, E. Blem, R. St. Amant, K. Sankaralingam, and D. Burger, "Dark silicon and the end of multicore scaling," in *Proceedings of the International Symposium on Computer Architecture*, 2011, pp. 365–376. DOI: 10.1145/2024723.2000108. 5, 6, 32

[55] H. Esmaeilzadeh, A. Sampson, L. Ceze, and D. Burger, "Architecture support for disciplined approximate programming," in *Proceedings of the Architectural Support for Programming Languages and Operating Systems*, Mar. 2012, pp. 301–312. DOI: 10.1145/2150976.2151008. 42

[56] S. Eyerman and L. Eeckhout, "A counter architecture for online DVFS profitability estimation," *IEEE Transactions on Computers*, vol. 59, no. 11, pp. 1576–1583, 2010. DOI: 10.1109/TC.2010.65. 17, 20, 21

[57] S. Eyerman, L. Eeckhout, T. Karkhanis, and J. E. Smith, "A performance counter architecture for computing accurate CPI components," *ACM SIGOPS Operating Systems Review*, vol. 40, no. 5, pp. 175–184, 2006. DOI: 10.1145/1168857.1168880. 17

[58] K. Flautner, N. S. Kim, S. Martin, D. Blaauw, and T. Mudge, "Drowsy caches: simple techniques for reducing leakage power," in *Proceedings of the International Symposium on Computer Architecture*, May 2002, pp. 148–157. DOI: 10.1109/ISCA.2002.1003572. 53

[59] K. Flautner, S. Reinhardt, and T. Mudge, "Automatic performance setting for dynamic voltage scaling," *Wireless Networks*, vol. 8, no. 5, pp. 507–520, Sep. 2002. DOI: 10.1023/A:1016546330128. 23

[60] Gartner, "Gartner says annual smartphone sales surpassed sales of feature phones for the first time in 2013," 2014, Gartner press release. [Online]. Available: http://www.gartner.com/newsroom/id/2665715 60

[61] P. Gavin, D. Whalley, and M. Själander, "Reducing instruction fetch energy in multi-issue processors," *ACM Transactions on Architecture and Code Optimization*, vol. 10, no. 4, pp. 64:1–64:24, Dec. 2013. DOI: 10.1145/2541228.2555318. 51

[62] M. Gebhart, S. W. Keckler, and W. J. Dally, "A compile-time managed multi-level register file hierarchy," in *Proceedings of the IEEE/ACM Annual International Symposium on Microarchitecture*, Dec. 2011, pp. 465–476. DOI: 10.1145/2155620.2155675. 57

[63] M. Gebhart, S. W. Keckler, B. Khailany, R. Krashinsky, and W. J. Dally, "Unifying primary cache, scratch, and register file memories in a throughput processor," in *Proceedings of the IEEE/ACM Annual International Symposium on Microarchitecture*, Dec. 2012, pp. 96–106. DOI: 10.1109/MICRO.2012.18. 56

[64] B. Goel, S. A. McKee, and M. Själander, *Techniques to Measurement, Model, and Manage Power*. Elsevier Advances in Computers, Green and Sustainable Computing: Part I, Nov. 2012, vol. 87. [Online]. Available: http://goo.gl/FCD9f 21, 22

[65] B. Goel, S. McKee, R. Gioiosa, K. Singh, M. Bhadauria, and M. Cesati, "Portable, scalable, per-core power estimation for intelligent resource management," in *Proceedings of the International Green Computing Conference*, Aug. 2010, pp. 135–146. DOI: 10.1109/GREENCOMP.2010.5598313. 21, 22

[66] N. Goulding, J. Sampson, G. Venkatesh, S. Garcia, J. Auricchio, J. Babb, M. Taylor, and S. Swanson, "GreenDroid: A mobile application processor for a future of dark silicon," 2010, HotChips presentation. [Online]. Available: http://www.hotchips.org/wp-content/uploads/hc_archives/hc22/HC22.23.240-Goulding-GreenDroid.pdf 32

[67] N. Goulding-Hotta, J. Sampson, G. Venkatesh, S. Garcia, J. Auricchio, P. Huang, M. Arora, S. Nath, V. Bhatt, J. Babb, S. Swanson, and M. Taylor, "The GreenDroid mobile application processor: An architecture for silicon's dark future," *IEEE Micro*, pp. 86–95, Mar. 2011. DOI: 10.1109/MM.2011.18.

[68] N. Goulding-Hotta, J. Sampson, Q. Zheng, V. Bhatt, S. Swanson, and M. Taylor, "GreenDroid: An architecture for the dark silicon age," in *Proceedings of the Asia and South Pacific Design Automation Conference*, 2012, pp. 100–105. DOI: 10.1109/ASP-DAC.2012.6164926. 32

[69] *The Green500 List*. [Online]. Available: http://www.green500.org 39

[70] P. Greenhalgh, "Big.LITTLE processing with ARM Cortex-A15 and Cortex-A7," 2011, ARM white paper. [Online]. Available: http://www.arm.com/files/downloads/big_LITTLE_Final_Final.pdf 2, 34, 35, 36

[71] E. Grochowski and M. Annavaram, "Energy per instruction trends in intel microprocessors," *Technology@ Intel Magazine*, vol. 4, no. 3, pp. 1–8, 2006. [Online]. Available: http://www.intel.com/pressroom/kits/core2duo/pdf/epi-trends-final2.pdf 3

[72] P. Hammarlund, "4th generation Intel® Core™ processor, codenamed Haswell," Aug. 2013, HotChips presentation. [Online]. Available: http://www.hotchips.org/wp-content/uploads/hc_archives/hc25/HC25.80-Processors2-epub/HC25.27.820-Haswell-Hammarlund-Intel.pdf 46

[73] R. Hegde and N. R. Shanbhag, "Energy-efficient signal processing via algorithmic noise-tolerance," in *Proceedings of the International Symposium on Low Power Electronics and Design*, 1999, pp. 30–35. DOI: 10.1145/313817.313834. 42

[74] S. Hines, Y. Peress, P. Gavin, D. Whalley, and G. Tyson, "Guaranteeing instruction fetch behavior with a lookahead instruction fetch engine (LIFE)," in *Proceedings of the ACM SIGPLAN/SIGBED Conference on Languages, Compilers, and Tools for Embedded Systems*, Jun. 2009, pp. 119–128. DOI: 10.1145/1542452.1542469. 51

[75] D. Hisamoto, W.-C. Lee, J. Kedzierski, H. Takeuchi, K. Asano, C. Kuo, E. Anderson, T.-J. King, J. Bokor, and C. Hu, "FinFET-a self-aligned double-gate MOSFET scalable to 20 nm," *IEEE Transactions on Electron Devices*, vol. 47, no. 12, pp. 2320–2325, Dec. 2000. DOI: 10.1109/16.887014. 14

[76] S. Hong and H. Kim, "An integrated GPU power and performance model," in *Proceedings of the International Symposium on Computer Architecture*, 2010, pp. 280–289. DOI: 10.1145/1816038.1815998. 9

[77] M. Horowitz, "Computing's energy problem (and what we can do about it)," in *Proceedings of the IEEE International Solid-State Circuits Conference*, Feb. 2014, pp. 10–14. DOI: 10.1109/ISSCC.2014.6757323. 50

[78] M. Horowitz, T. Indermaur, and R. Gonzalez, "Low-power digital design," in *Proceedings of the IEEE Symposium on Low Power Electronics*, 1994, pp. 8–11. DOI: 10.1109/LPE.1994.573184. 11

[79] Y. Hoskote, S. Vangal, A. Singh, N. Borkar, and S. Borkar, "A 5-GHz mesh interconnect for a teraflops processor," *IEEE Micro*, vol. 27, no. 5, pp. 51–61, Sep. 2007. DOI: 10.1109/MM.2007.4378783. 47

[80] J. Howard, "A 48-core IA-32 processor with on-die message-passing and DVFS in 45nm CMOS," in *Proceedings of the IEEE International Solid-State Circuits Conference*, 2010, pp. 1–4. DOI: 10.1109/ISSCC.2010.5434077. 47

[81] J. Howard, S. Dighe, Y. Hoskote, S. Vangal, D. Finan, G. Ruhl, D. Jenkins, H. Wilson, N. Borkar, G. Schrom, F. Pailet, S. Jain, T. Jacob, S. Yada, S. Marella, P. Salihundam, V. Erraguntla, M. Konow, M. Riepen, G. Droege, J. Lindemann, M. Gries, T. Apel, K. Henriss, T. Lund-Larsen, S. Steibl, S. Borkar, V. De, R. Van der Wijngaart, and T. Mattson, "A 48-core IA-32 message-passing processor with DVFS in 45nm CMOS," in *Proceedings of the IEEE International Solid-State Circuits Conference*, 2010, pp. 108–109. DOI: 10.1109/ISSCC.2010.5434077. 47

[82] C.-H. Hsu and U. Kremer, "The design, implementation, and evaluation of a compiler algorithm for CPU energy reduction," in *Proceedings of the ACM SIGPLAN Conference on Programming Language Design and Implementation*, 2003, pp. 38–48. DOI: 10.1145/780822.781137. 11, 17

[83] Imagination, "PowerVR Series6XT," Feb. 2013, web page. [Online]. Available: http://www.imgtec.com/powervr/series6xt.asp 40

[84] K. Inoue, T. Ishihara, and K. Murakami, "Way-predicting set-associative cache for high performance and low energy consumption," in *Proceedings of the International Symposium on Low Power Electronics and Design*, 1999, pp. 273–275. DOI: 10.1145/313817.313948. 51

[85] Intel, "Enhanced Intel® SpeedStep® technology for the Intel® Pentium® M processor," Mar. 2004, Intel white paper. [Online]. Available: http://download.intel.com/design/network/papers/30117401.pdf 22

[86] Intel, "Intel® Turbo Boost technology in Intel® Core™ microarchitecture (Nehalem) based processors," Nov. 2008, Intel white paper. [Online]. Available: http://files.shareholder.com/downloads/INTC/0x0x348508/C9259E98-BE06-42C8-A433-E28F64CB8EF2/TurboBoostWhitePaper.pdf 17, 34

[87] Intel, "4th generation Intel® Core™ processor-based platforms for intelligent systems," 2013, Intel platform brief. [Online]. Available: http://www.intel.com/content/dam/www/public/us/en/documents/platform-briefs/core-q87-chipset-is-brief.pdf 39

[88] Intel, "New microarchitecture for 4th gen Intel® Core™ processor platforms," 2013, Intel product brief. [Online]. Available: http://www.intel.se/content/dam/www/public/us/en/documents/product-briefs/4th-gen-core-family-mobile-brief.pdf 15, 47

[89] Intel, "Disrupting the data center to create the digital services economy," Jun. 2014, Intel blog post in The Data Stack. [Online]. Available: https://communities.intel.com/community/itpeernetwork/datastack/blog/2014/06/18/disrupting-the-data-center-to-create-the-digital-services-economy 42

[90] E. Ipek, M. Kirman, N. Kirman, and J. F. Martinez, "Core fusion: Accommodating software diversity in chip multiprocessors," in *Proceedings of the International Symposium on Computer Architecture*, 2007, pp. 186–197. DOI: 10.1145/1273440.1250686. 38

[91] IRC, "The international technology roadmap for semiconductors (ITRS)," 2013, IRC overview. [Online]. Available: http://www.itrs.net/Links/2013ITRS/2013Chapters/2013Overview.pdf 1, 6

[92] C. Isci, G. Contreras, and M. Martonosi, "Live, runtime phase monitoring and prediction on real systems with application to dynamic power management," in *Proceedings of the IEEE/ACM Annual International Symposium on Microarchitecture*, 2006, pp. 359–370. DOI: 10.1109/MICRO.2006.30. 11, 17

[93] C. Isci and M. Martonosi, "Identifying program power phase behavior using power vectors," in *Proceedings of the IEEE International Workshop on Workload Characterization*, 2003, pp. 108–118. DOI: 10.1109/WWC.2003.1249062. 11, 17

[94] B. Jacob, "The memory system: You can't avoid it, you can't ignore it, you can't fake it," *Synthesis Lectures on Computer Architecture*, vol. 4, no. 1, pp. 1–77, 2009. DOI: 10.2200/S00201ED1V01Y200907CAC007. 45

[95] S. Jahagirdar, V. George, I. Sodhi, and R. Wells, "Power management of the third generation Intel Core micro architecture formerly codenamed Ivy Bridge," Aug. 2012, HotChips presentation. http://www.hotchips.org/wp-content/uploads/hc_archives/hc24/HC24-1-Microprocessor/HC24.28.117-HotChips_IvyBridge_Power_04.pdf 52, 53

[96] N. E. Jerger and L.-S. Peh, "On-chip networks," *Synthesis Lectures on Computer Architecture*, vol. 4, no. 1, pp. 1–141, 2009. DOI: 10.2200/S00209ED1V01Y200907CAC008. 45

[97] N. E. Jerger, L.-S. Peh, and M. Lipasti, "Virtual circuit tree multicasting: A case for on-chip hardware multicast support," in *Proceedings of the International Symposium on Computer Architecture*, 2008, pp. 229–240. DOI: 10.1145/1394608.1382141. 48

[98] Y. Jin, K. H. Yum, and E. J. Kim, "Adaptive data compression for high-performance low-power on-chip networks," in *Proceedings of the IEEE/ACM Annual International Symposium on Microarchitecture*, Nov. 2008, pp. 354–363. DOI: 10.1109/MICRO.2008.4771804. 48

[99] N. P. Jouppi, P. Boyle, and J. S. Fitch, "Designing, packaging, and testing a 300-MHz, 115 W ECL microprocessor," *IEEE Micro*, vol. 14, no. 2, pp. 50–58, Apr. 1994. DOI: 10.1109/40.272837. 1, 59

[100] N. Jouppi, P. Boyle, J. Dion, M. Doherty, A. Eustace, R. Haddad, R. Mayo, S. Menon, L. Monier, D. Stark, S. Turrini, J. Yang, W. Hamburgen, J. Fitch, and R. Kao, "A 300-MHz 115-W 32-b bipolar ECL microprocessor," *IEEE Journal of Solid-State Circuits*, vol. 28, no. 11, pp. 1152–1166, 1993. DOI: 10.1109/4.245601. 1

[101] A. Kahng, S. Kang, R. Kumar, and J. Sartori, "Designing processors from the ground up to allow voltage/reliability tradeoffs," in *Proceedings of the International Symposium High-Performance Computer Architecture*, Jan. 2010, pp. 1–11. DOI: 10.1109/HPCA.2010.5416652. 42

[102] T. S. Karkhanis and J. E. Smith, "A first-order superscalar processor model," in *ACM SIGARCH Computer Architecture News*, vol. 32, no. 2, 2004, pp. 338–349. DOI: 10.1145/1028176.1006729. 17

[103] S. Kaxiras and M. Martonosi, "Computer architecture techniques for power-efficiency," *Synthesis Lectures on Computer Architecture*, vol. 3, no. 1, pp. 1–207, 2008. DOI: 10.2200/S00119ED1V01Y200805CAC004. ix, 9, 11, 17, 59, 60

[104] S. Keckler, "Life after Dennard and how I learned to love the picojoule," 2011, keynote at MICRO. 39, 45, 46

[105] C. N. Keltcher, "The AMD Hammer processor core," Aug. 2002, HotChips presentation. [Online]. Available: http://www.hotchips.org/wp-content/uploads/hc_archives/hc14/3_Tue/27_AMD_Hammer_Core_HC_v5.pdf 16, 57

[106] G. Keramidas, V. Spiliopoulos, and S. Kaxiras, "Interval-based models for run-time DVFS orchestration in superscalar processors," in *Proceedings of the ACM International Conference on Computing Frontiers*, 2010, pp. 287–296. DOI: 10.1145/1787275.1787338. 17, 18, 19, 20, 21

[107] R. E. Kessler, "The Cavium 32 core OCTEON II 68xx," Aug. 2011, HotChips presentation. [Online]. Available: http://www.hotchips.org/wp-content/uploads/hc_archives/hc23/HC23.18.1-manycore/HC23.18.110-OCTEON_II-Kessler-Cavium%20-%20Copy.pdf 47

[108] R. E. Kessler, "Multicore and the 32 core Cavium OCTEON II 68xx," Feb. 2013, seminar slides. [Online]. Available: http://www.ece.cmu.edu/~calcm/lib/exe/fetch.php?media=seminars:2013_spring_cavium.pdf 47

[109] C. Kim, S. Sethumadhavan, M. S. Govindan, N. Ranganathan, D. Gulati, D. Burger, and S. W. Keckler, "Composable lightweight processors," in *Proceedings of the IEEE/ACM Annual International Symposium on Microarchitecture*, Dec. 2007, pp. 381–394. 37, 38

[110] W. Kim, M. S. Gupta, G.-Y. Wei, and D. Brooks, "System level analysis of fast, per-core DVFS using on-chip switching regulators," in *Proceedings of the International Symposium High-Performance Computer Architecture*, 2008, pp. 123–134. DOI: 10.1109/HPCA.2008.4658633. 15

[111] J. Kin, M. Gupta, and W. H. Mangione-Smith, "The filter cache: An energy efficient memory structure," in *Proceedings of the IEEE/ACM Annual International Symposium on Microarchitecture*, Dec. 1997, pp. 184–193. DOI: 10.1109/MICRO.1997.645809. 50

[112] J. Kin, M. Gupta, and W. H. Mangione-Smith, "Filtering memory references to increase energy efficiency," *IEEE Transactions on Computers*, vol. 49, no. 1, pp. 1–15, Jan. 2000. DOI: 10.1109/12.822560. 50

[113] N. Kirman, M. Kirman, R. K. Dokania, J. F. Martinez, A. B. Apsel, M. A. Watkins, and D. H. Albonesi, "Leveraging optical technology in future bus-based chip multiprocessors," in *Proceedings of the IEEE/ACM Annual International Symposium on Microarchitecture*, 2006, pp. 492–503. DOI: 10.1109/MICRO.2006.28. 49

[114] N. Kirman and J. F. Martínez, "A power-efficient all-optical on-chip interconnect using wavelength-based oblivious routing," in *Proceedings of the Architectural Support for Programming Languages and Operating Systems*, 2010, pp. 15–28. DOI: 10.1145/1736020.1736024. 49

[115] B. Klug and A. L. Shimpi, "Qualcomm's new Snapdragon S4: MSM8960 & Krait architecture explored," Oct. 2011, web page. [Online]. Available: http://www.anandtech.com/show/4940/qualcomm-new-snapdragon-s4-msm8960-krait-architecture/2 50

[116] R. Kumar, D. Tullsen, P. Ranganathan, N. Jouppi, and K. Farkas, "Single-ISA heterogeneous multi-core architectures for multithreaded workload performance," in *Proceedings of the International Symposium on Computer Architecture*, Jun. 2004, pp. 64–75. DOI: 10.1145/1028176.1006707. 34, 35

[117] R. Kumar, D. M. Tullsen, N. P. Jouppi, and P. Ranganathan, "Heterogeneous chip multiprocessors," *Computer*, vol. 38, no. 11, pp. 32–38, Nov. 2005. DOI: 10.1109/MC.2005.379. 25

[118] G. Kyriazis, "Heterogeneous system architecture: A technical review," Aug. 2012, AMD white paper. [Online]. Available: http://developer.amd.com/wordpress/media/2012/10/hsa10.pdf 46

[119] HP Labs, "CACTI an integrated cache and memory access time, cycle time, area, leakage, and dynamic power model," HP Labs. [Online]. Available: http://www.hpl.hp.com/research/cacti/ 21

[120] L. Lee, B. Moyer, and J. Arends, "Instruction fetch energy reduction using loop caches for embedded applications with small tight loops," in *Proceedings of the International Symposium on Low Power Electronics and Design*, 1999, pp. 267–269. DOI: 10.1145/313817.313944. 50

[121] L. Leem, H. Cho, J. Bau, Q. A. Jacobson, and S. Mitra, "ERSA: Error resilient system architecture for probabilistic applications," in *Proceedings of the Conference on Design, Automation and Test in Europe*, 2010, pp. 1560–1565. DOI: 10.1109/DATE.2010.5457059. 42

[122] J. Leng, T. Hetherington, A. ElTantawy, S. Gilani, N. S. Kim, T. M. Aamodt, and V. J. Reddi, "GPUWattch: Enabling energy optimizations in GPGPUs," in *Proceed-

ings of the International Symposium on Computer Architecture, 2013, pp. 487–498. DOI: 10.1145/2485922.2485964. 9

[123] J. Li and J. F. Martínez, "Dynamic power-performance adaptation of parallel computation on chip multiprocessors," in *Proceedings of the International Symposium High-Performance Computer Architecture*, Feb. 2006, pp. 77–87. DOI: 10.1109/HPCA.2006.1598114. 25

[124] J. Li, J. F. Martinez, and M. C. Huang, "The thrifty barrier: Energy-aware synchronization in shared-memory multiprocessors," in *Proceedings of the International Symposium High-Performance Computer Architecture*, 2004, pp. 14–23. DOI: 10.1109/HPCA.2004.10018. 27

[125] L. Li, H. Feng, and J. Xue, "Compiler-directed scratchpad memory management via graph coloring," *ACM Transactions on Architecture and Code Optimization*, vol. 6, no. 3, pp. 9:1–9:17, Oct. 2009. DOI: 10.1145/1582710.1582711. 56

[126] L. Li, L. Gao, and J. Xue, "Memory coloring: A compiler approach for scratchpad memory management," in *Proceedings of the International Conference on Parallel Architectural and Compilation Techniques*, Aug. 2005, pp. 329–338. DOI: 10.1109/PACT.2005.27. 56

[127] S. Li, J. H. Ahn, J. B. Brockman, and N. P. Jouppi, "McPAT 1.0: An integrated power, area, and timing modeling framework for multicore architectures," *Proceedings of the IEEE/ACM Annual International Symposium on Microarchitecture*, pp. 469–480, Dec. 2010. DOI: 10.1145/1669112.1669172. 9, 21

[128] P. Lotfi-Kamran, M. Ferdman, D. Crisan, and B. Falsafi, "TurboTag: Lookup filtering to reduce coherence directory power," in *Proceedings of the International Symposium on Low Power Electronics and Design*, Aug. 2010, pp. 377–382. DOI: 10.1145/1840845.1840929. 56

[129] M. Lyons, G.-Y. Wei, and D. Brooks, "Shrink-fit: A framework for flexible accelerator sizing," *IEEE Computer Architecture Letters*, vol. 12, no. 1, pp. 17–20, 2013. DOI: 10.1109/L-CA.2012.7. 41

[130] M. J. Lyons, M. Hempstead, G.-Y. Wei, and D. Brooks, "The accelerator store: A shared memory framework for accelerator-based systems," *ACM Transactions on Architecture and Code Optimization*, vol. 8, no. 4, pp. 48:1–48:22, Jan. 2012. DOI: 10.1145/2086696.2086727. 41

[131] F. Mattern and C. Floerkemeier, "From the internet of computers to the internet of things," in *From active data management to event-based systems and more.* Springer, 2010, pp. 242–259. DOI: 10.1007/978-3-642-17226-7_15. 60

[132] R. Merritt, "ARM CTO: Power surge could create 'dark silicon'," Oct. 2009. [Online]. Available: http://www.eetimes.com/document.asp?doc_id=1172049 32

[133] R. Merritt, "Group describes specs for x86, ARM SoCs," 2013, web page. [Online]. Available: http://www.eetimes.com/document.asp?doc_id=1319306 7

[134] R. Miftakhutdinov, E. Ebrahimi, and Y. N. Patt, "Predicting performance impact of DVFS for realistic memory systems," in *Proceedings of the IEEE/ACM Annual International Symposium on Microarchitecture*, Dec. 2012, pp. 155–165. DOI: 10.1109/MICRO.2012.23. 21

[135] R. Mijat, "Take GPU processing power beyond graphics with Mali GPU computing," Aug. 2012, ARM white paper. [Online]. Available: http://malideveloper.arm.com/downloads/WhitePaper_GPU_Computing_on_Mali.pdf 39

[136] R. Mishra, N. Rastogi, D. Zhu, D. Mossé, and R. Melhem, "Energy aware scheduling for distributed real-time systems," in *Proceedings of the IEEE International Parallel and Distributed Processing Symposium*, Apr. 2003. DOI: 10.1109/IPDPS.2003.1213099. 26, 29

[137] C. Moore, "Framework for innovation," 2007, FCRC/ISCA Conference, San Diego, CA. Invited keynote. 5

[138] G. E. Moore, "Cramming more components onto integrated circuits," *Electronics*, vol. 38, no. 8, pp. 114–117, Apr. 1998. DOI: 10.1109/JPROC.1998.658762. 1

[139] T. Y. Morad, U. C. Weiser, A. Kolodny, M. Valero, and E. Ayguade, "Performance, power efficiency and scalability of asymmetric cluster chip multiprocessors," *IEEE Computer Architecture Letters*, vol. 5, no. 1, pp. 4–17, Jan. 2006. DOI: 10.1109/L-CA.2006.6. 25

[140] A. Moshovos, "RegionScout: Exploiting coarse grain sharing in snoop-based coherence," in *Proceedings of the International Symposium on Computer Architecture*, May 2005, pp. 234–245. DOI: 10.1145/1080695.1069990. 55

[141] A. Moshovos, G. Memik, A. Choudhary, and B. Falsafi, "JETTY: Filtering snoops for reduced energy consumption in SMP servers," in *Proceedings of the International Symposium High-Performance Computer Architecture*, Jan. 2001, pp. 85–96. DOI: 10.1109/HPCA.2001.903254. 54

[142] S. S. Mukherjee, C. T. Weaver, J. Emer, S. K. Reinhardt, and T. Austin, "Measuring architectural vulnerability factors," *Proceedings of the IEEE/ACM Annual International Symposium on Microarchitecture*, vol. 23, pp. 70–75, 2003. DOI: 10.1109/MM.2003.1261389. 42

[143] M. Muller, "ARM keynote: Will 'Dark Silicon' derail the mobile internet?" EE Times "Designing with ARM" virtual conference, Mar. 2010. [Online]. Available: http://www.cadence.com/Community/blogs/ii/archive/2010/03/31/arm-keynote-will-dark-silicon-derail-the-mobile-internet.aspx 7

[144] NVIDIA, "NVIDIA Tegra 4 family CPU architecture," NVIDIA white paper. [Online]. Available: http://www.nvidia.com/docs/IO/116757/NVIDIA_Quad_a15_whitepaper_FINALv2.pdf 7, 36

[145] NVIDIA, "Variable SMP multi-core CPU architecture for low power and high performance," NVIDIA white paper. [Online]. Available: http://www.nvidia.com/content/PDF/tegra_white_papers/tegra-whitepaper-0911b.pdf 36

[146] NVIDIA, "GeForce GTX 770 specification," 2013, web page. [Online]. Available: http://www.geforce.com/hardware/desktop-gpus/geforce-gtx-770/specifications 39

[147] S.-H. Oh, "Physics and technologies of vertical transistors," Ph.D. dissertation, Standford University, Jun. 2001. 13

[148] Oracle, "Oracle's SPARC T5-2, SPARC T5-4, SPARC T5-8, and SPARC T5-1B server architecture," Feb. 2014, Oracle white paper. [Online]. Available: http://www.oracle.com/technetwork/server-storage/sun-sparc-enterprise/documentation/o13-024-sparc-t5-architecture-1920540.pdf 47

[149] J. A. Paradiso and T. Starner, "Energy scavenging for mobile and wireless electronics," *IEEE Pervasive Computing*, vol. 4, no. 1, pp. 18–27, 2005. DOI: 10.1109/MPRV.2005.9. 60

[150] J. Park, D. Shin, N. Chang, and M. Pedram, "Accurate modeling and calculation of delay and energy overheads of dynamic voltage scaling in modern high-performance microprocessors," in *Proceedings of the International Symposium on Low Power Electronics and Design*, Aug. 2010, pp. 419–424. DOI: 10.1145/1840845.1840938. 15

[151] S. Park, T. Krishna, C.-H. Chen, B. Daya, A. Chandrakasan, and L.-S. Peh, "Approaching the theoretical limits of a mesh NoC with a 16-node chip prototype in 45nm SOI," in *Proceedings of the ACM/IEEE Design Automation Conference*, 2012, pp. 398–405. DOI: 10.1145/2228360.2228431. 48

[152] A. Pathak, Y. C. Hu, and M. Zhang, "Where is the energy spent inside my app?: Fine grained energy accounting on smartphones with eprof," in *Proceedings of the ACM European Conference on Computer Systems*, 2012, pp. 29–42. DOI: 10.1145/2168836.2168841. 1

[153] N. Pinckney, R. Dreslinski, K. Sewell, D. Fick, T. Mudge, D. Sylvester, and D. Blaauw, "Limits of parallelism and boosting in dim silicon," *IEEE Micro*, vol. 33, no. 5, pp. 30–37, Sep. 2013. DOI: 10.1109/MM.2013.73. 33

[154] F. J. Pollack, "New microarchitecture challenges in the coming generations of CMOS process technologies," in *Proceedings of the IEEE/ACM Annual International Symposium on Microarchitecture*, Nov. 1999, p. 2, keynote. 4

[155] J. Postman, T. Krishna, C. Edmonds, L.-S. Peh, and P. Chiang, "SWIFT: A low-power network-on-chip implementing the token flow control router architecture with swing-reduced interconnects," *IEEE Transactions on Very Large Scale Integration (VLSI) Systems*, vol. 21, no. 8, pp. 1432–1446, 2013. DOI: 10.1109/TVLSI.2012.2211904. 48

[156] M. D. Powell, A. Agarwal, T. N. Vijaykumar, B. Falsafi, and K. Roy, "Reducing set-associative cache energy via way-prediction and selective direct-mapping," in *Proceedings of the IEEE/ACM Annual International Symposium on Microarchitecture*, 2001, pp. 54–65. DOI: 10.1109/MICRO.2001.991105. 51

[157] P. Pujara and A. Aggarwal, "Cache noise prediction," *IEEE Transactions on Computers*, vol. 57, no. 10, pp. 1372–1386, Oct. 2008. DOI: 10.1109/TC.2008.75. 51

[158] A. Putnam, A. Caulfield, E. Chung, D. Chiou, K. Constantinides, J. Demme, H. Esmaeilzadeh, J. Fowers, G. Gopal, J. Gray, M. Haselman, S. Hauck, S. Heil, A. Hormati, J.-Y. Kim, S. Lanka, J. Larus, E. Peterson, S. Pope, A. Smith, J. Thong, P. Xiao, and D. Burger, "A reconfigurable fabric for accelerating large-scale datacenter services," in *Proceedings of the International Symposium on Computer Architecture*, Jun. 2014, pp. 13–24. DOI: 10.1145/2678373.2665678. 41

[159] Qualcomm, "Qualcomm© Snapdragon 800 processors," Jan. 2014, Qualcomm product brief. [Online]. Available: http://www.qualcomm.com/media/documents/files/qualcomm-snapdragon-800-product-brief.pdf 7

[160] B. Rountree, D. K. Lowenthal, M. Schulz, and B. R. de Supinski, "Practical performance prediction under dynamic voltage frequency scaling," in *Proceedings of the International Green Computing Conference*, 2011, pp. 1–8. DOI: 10.1109/IGCC.2011.6008553. 17

[161] V. Salapura, M. Blumrich, and A. Gara, "Design and implementation of the Blue Gene/P snoop filter," in *Proceedings of the International Symposium High-Performance Computer Architecture*, Feb. 2008, pp. 5–14. DOI: 10.1109/HPCA.2008.4658623. 55

[162] V. Salapura, M. Blumrich, and A. Gara, "Improving the accuracy of snoop filtering using stream registers," in *Proceedings of the Workshop on MEmory Performance: DEaling with Applications, Systems and Architecture*, Sep. 2007, pp. 25–32. DOI: 10.1145/1327171.1327174. 55

[163] A. Sampson, W. Dietl, E. Fortuna, D. Gnanapragasam, L. Ceze, and D. Grossman, "EnerJ: Approximate data types for safe and general low-power computation," in *Proceedings*

of the ACM SIGPLAN Conference on Programming Language Design and Implementation, Jun. 2011, pp. 164–174. DOI: 10.1145/1993316.1993518. 42, 43

[164] H. Saputra, M. Kandemir, N. Vijaykrishnan, M. Irwin, J. Hu, and U. Kremer, "Energy-conscious compilation based on voltage scaling," in *Proceedings of the ACM SIGPLAN Joint Conference on Languages, Compilers, and Tools for Embedded Systems and Software and Compilers for Embedded Systems*, 2002, pp. 2–11. DOI: 10.1145/566225.513832. 11, 17

[165] S. Satpathy, K. Sewell, T. Manville, Y.-P. Chen, R. Dreslinski, D. Sylvester, T. Mudge, and D. Blaauw, "A 4.5Tb/s 3.4Tb/s/W 64x64 switch fabric with self-updating least-recently-granted priority and quality-of-service arbitration in 45nm CMOS," in *Proceedings of the IEEE International Solid-State Circuits Conference*, 2012, pp. 478–480. DOI: 10.1109/ISSCC.2012.6177098. 48

[166] L. Seiler, D. Carmean, E. Sprangle, T. Forsyth, M. Abrash, P. Dubey, S. Junkins, A. Lake, J. Sugerman, R. Cavin, R. Espasa, E. Grochowski, T. Juan, and P. Hanrahan, "Larrabee: a many-core x86 architecture for visual computing," in *Proceedings of SIGGRAPH: International Conference on Computer Graphics and Interactive Techniques*, 2008, pp. 18:1–18:15. DOI: 10.1145/1360612.1360617. 47

[167] A. Sembrant, D. Eklov, and E. Hagersten, "Efficient software-based online phase classification," in *Proceedings of the IEEE International Symposium on Workload Characterization*, 2011, pp. 104–115. DOI: 10.1109/IISWC.2011.6114207. 22

[168] G. Semeraro, D. H. Albonesi, S. G. Dropsho, G. Magklis, S. Dwarkadas, and M. L. Scott, "Dynamic frequency and voltage control for a multiple clock domain microarchitecture," in *Proceedings of the IEEE/ACM Annual International Symposium on Microarchitecture*, 2002, pp. 356–367. DOI: 10.1109/MICRO.2002.1176263. 16

[169] G. Semeraro, G. Magklis, R. Balasubramonian, D. H. Albonesi, S. Dwarkadas, and M. L. Scott, "Energy-efficient processor design using multiple clock domains with dynamic voltage and frequency scaling," in *Proceedings of the International Symposium High-Performance Computer Architecture*, 2002, pp. 29–40. DOI: 10.1109/HPCA.2002.995696. 16

[170] K. Sewell, R. Dreslinski, T. Manville, S. Satpathy, N. Pinckney, G. Blake, M. Cieslak, R. Das, T. Wenisch, D. Sylvester, D. Blaauw, and T. Mudge, "Swizzle-switch networks for many-core systems," *IEEE Transactions on Emerging and Selected Topics in Circuits and Systems*, vol. 2, no. 2, pp. 278–294, 2012. DOI: 10.1109/JETCAS.2012.2193936. 48

[171] A. Shacham, K. Bergman, and L. Carloni, "Photonic networks-on-chip for future generations of chip multiprocessors," *IEEE Transactions on Computers*, vol. 57, no. 9, pp. 1246–1260, 2008. DOI: 10.1109/TC.2008.78. 49

[172] L. Shang, L.-S. Peh, and N. K. Jha, "Dynamic voltage scaling with links for power optimization of interconnection networks," in *Proceedings of the International Symposium High-Performance Computer Architecture*, pp. 91-102, Feb. 2003. DOI: 10.1109/HPCA.2003.1183527. 47

[173] A. L. Shimpi, "Intel's Haswell architecture analyzed: Building a new PC and a new Intel," 2013, web page. [Online]. Available: http://www.anandtech.com/show/6355/intels-haswell-architecture/10 46

[174] Y. Shin, K. Shin, P. Kenkare, R. Kashyap, H.-J. Lee, D. Seo, B. Millar, Y. Kwon, R. Iyengar, M.-S. Kim, A. Chowdhury, S.-I. Bae, I. Hong, W. Jeong, A. Lindner, U. Cho, K. Hawkins, J. C. Son, and S. H. Hwang, "28nm high- metal-gate heterogeneous quad-core CPUs for high-performance and energy-efficient mobile application processor," in *Proceedings of the IEEE International Solid-State Circuits Conference*, 2013, pp. 154–155. DOI: 10.1109/ISSCC.2013.6487678. 36

[175] B. Sinharoy, R. Kalla, W. J. Starke, H. Q. Le, R. Cargnoni, J. A. Van Norstrand, B. J. Ronchetti, J. Stuecheli, J. Leenstra, G. L. Guthrie, D. Q. Nguyen, B. Blaner, C. F. Marino, E. Retter, and P. Williams, "IBM POWER7 multicore server processor," *IBM Journal of Research and Development*, vol. 55, no. 3, pp. 191–219, May 2011. DOI: 10.1147/JRD.2011.2127330. 46

[176] V. Soteriou and L.-S. Peh, "Exploring the design space of self-regulating power-aware on/off interconnection networks," *IEEE Transactions on Parallel and Distributed Systems*, vol. 18, no. 3, pp. 393–408, 2007. DOI: 10.1109/TPDS.2007.43. 48

[177] V. Spiliopoulos, S. Kaxiras, and G. Keramidas, "Green governors: A framework for continuously adaptive DVFS," in *Proceedings of the International Green Computing Conference*, Jul. 2011, pp. 1–8. DOI: 10.1109/IGCC.2011.6008552. 21, 24

[178] V. Spiliopoulos, A. Sembrant, and S. Kaxiras, "Power-sleuth: A tool for investigating your program's power behavior," in *Proceedings of the IEEE International Symposium on Modeling, Analysis Simulation of Computer and Telecommunication Systems*, 2012, pp. 241–250. DOI: 10.1109/MASCOTS.2012.36. 22

[179] R. St. Amant *et al.*, "General-purpose code acceleration with limited-precision analog computation," in *Proceedings of the International Symposium on Computer Architecture*, pp. 505–516, Jun. 2014. DOI: 10.1145/2678373.2665746. 43

[180] I.-J. Sung, J. A. Stratton, and W.-M. W. Hwu, "Data layout transformation exploiting memory-level parallelism in structured grid many-core applications," in *Proceedings of the International Conference on Parallel Architectural and Compilation Techniques*, pp. 513–522, Sep. 2010. DOI: 10.1145/1854273.1854336. 56

[181] E. Talpes and D. Marculescu, "Toward a multiple clock/voltage island design style for power-aware processors," *IEEE Transactions on Very Large Scale Integration (VLSI) Systems*, vol. 13, no. 5, pp. 591–603, May 2005. DOI: 10.1109/TVLSI.2005.844305. 16

[182] M. Taylor, "Is dark silicon useful? Harnessing the four horsemen of the coming dark silicon apocalypse," in *Proceedings of the ACM/IEEE Design Automation Conference*, Jun. 2012, pp. 1131–1136. DOI: 10.1145/2228360.2228567 33

[183] TILERA, "TILE-Gx8072 processor," Jan. 2014, TILERA product brief. [Online]. Available: http://www.tilera.com/sites/default/files/images/products/TILE-Gx8072_PB 041-03_WEB.pdf 7

[184] K. Van Craeynest, A. Jaleel, L. Eeckhout, P. Narvaez, and J. Emer, "Scheduling heterogeneous multi-cores through performance impact estimation (PIE)," in *Proceedings of the International Symposium on Computer Architecture*, 2012, pp. 213–224. DOI: 10.1145/2366231.2337184. 37

[185] S. Venkataramani, V. K. Chippa, S. T. Chakradhar, K. Roy, and A. Raghunathan, "Quality programmable vector processors for approximate computing," in *Proceedings of the IEEE/ACM Annual International Symposium on Microarchitecture*, Dec. 2013, pp. 1–12. DOI: 10.1145/2540708.2540710. 42

[186] G. Venkatesh, J. Sampson, N. Goulding, S. Garcia, V. Bryksin, J. Lugo-Martinez, S. Swanson, and M. B. Taylor, "Conservation cores: reducing the energy of mature computations," in *Proceedings of the Architectural Support for Programming Languages and Operating Systems*, pp. 205–218, Mar. 2010. DOI: 10.1145/1736020.1736044. 5, 32

[187] M. Verma, L. Wehmeyer, and P. Marwedel, "Cache-aware scratchpad allocation algorithm," in *Proceedings of the Conference on Design, Automation and Test in Europe*, pp. 1264–1269, Jul. 2004. DOI: 10.1109/DATE.2004.1269069. 56

[188] M. Verma, L. Wehmeyer, and P. Marwedel, "Dynamic overlay of scratchpad memory for energy minimization," in *Proceedings of the ACM/IEEE International Conference on Hardware/Software Codesign and System Synthesis*, Sep. 2004, pp. 104–109. DOI: 10.1145/1016720.1016748. 56

[189] N. Verma, A. Shoeb, J. Bohorquez, J. Dawson, J. Guttag, and A. P. Chandrakasan, "A micro-power EEG acquisition SoC with integrated feature extraction processor for a chronic seizure detection system," *IEEE Journal of Solid-State Circuits*, vol. 45, no. 4, pp. 804–816, 2010. DOI: 10.1109/JSSC.2010.2042245. 60

[190] H. Wang, V. Sathish, R. Singh, M. J. Schulte, and N. S. Kim, "Workload and power budget partitioning for single-chip heterogeneous processors," in *Proceedings of the International*

Conference on Parallel Architectural and Compilation Techniques, 2012, pp. 401–410. DOI: 10.1145/2370816.2370873. 40

[191] War Department, "Physical aspects, operation of ENIAC are described," 1946. [Online]. Available: http://americanhistory.si.edu/comphist/pr4.pdf 1

[192] Y. Watanabe, J. D. Davis, and D. A. Wood, "WiDGET: Wisconsin decoupled grid execution tiles," in *Proceedings of the International Symposium on Computer Architecture*, Jun. 2010, pp. 2–13. DOI: 10.1145/1816038.1815965. 13, 37, 38

[193] M. Weiser, B. Welch, A. Demers, and S. Shenker, "Scheduling for reduced CPU energy," *Proceedings of the USENIX Conference on Operating System Design and Implementation*, pp. 13–23, 1994. DOI: 10.1007/978-0-585-29603-6_17. 22

[194] D. Wentzlaff, P. Griffin, H. Hoffmann, L. Bao, B. Edwards, C. Ramey, M. Mattina, C.-C. Miao, J. F. Brown III, and A. Agarwal, "On-chip interconnection architecture of the tile processor," *IEEE Micro*, vol. 27, no. 5, pp. 15–31, Sep. 2007. DOI: 10.1109/MM.2007.89. 47

[195] C. Wilkerson, H. Gao, A. Alameldeen, Z. Chishti, M. Khellah, and S.-L. Lu, "Trading off cache capacity for low-voltage operation," *IEEE Micro*, vol. 29, no. 1, pp. 96–103, Jan. 2009. DOI: 10.1109/MM.2009.20. 53

[196] C. Wilkerson, H. Gao, A. R. Alameldeen, Z. Chishti, M. Khellah, and S.-L. Lu, "Trading off cache capacity for reliability to enable low voltage operation," in *Proceedings of the International Symposium on Computer Architecture*, Jun. 2008, pp. 203–214. DOI: 10.1145/1394608.1382139. 53

[197] Q. Wu, V. Reddi, Y. Wu, J. Lee, D. Connors, D. Brooks, M. Martonosi, and D. Clark, "A dynamic compilation framework for controlling microprocessor energy and performance," in *Proceedings of the IEEE/ACM Annual International Symposium on Microarchitecture*, pp. 271–282, Nov. 2005. DOI: 10.1109/MICRO.2005.7. 11, 17

[198] Q. Wu, P. Juang, M. Martonosi, and D. W. Clark, "Formal online methods for voltage/frequency control in multiple clock domain microprocessors," in *Proceedings of the Architectural Support for Programming Languages and Operating Systems*, 2004, pp. 248–259. DOI: 10.1145/1024393.1024423. 16

[199] Q. Wu, M. Martonosi, D. W. Clark, V. J. Reddi, D. Connors, Y. Wu, J. Lee, and D. Brooks, "Dynamic-compiler-driven control for microprocessor energy and performance," *IEEE Micro*, vol. 26, no. 1, pp. 119–129, Jan. 2006. DOI: 10.1109/MM.2006.9. 16

[200] F. Xie, M. Martonosi, and S. Malik, "Compile-time dynamic voltage scaling settings: opportunities and limits," in *Proceedings of the ACM SIGPLAN Conference on Programming Language Design and Implementation*, 2003, pp. 49–62. DOI: 10.1145/781131.781138. 11, 17

[201] F. Xie, M. Martonosi, and S. Malik, "Intraprogram dynamic voltage scaling: Bounding opportunities with analytic modeling," *ACM Transactions on Architecture and Code Optimization*, vol. 1, no. 3, pp. 323–367, Sep. 2004. DOI: 10.1145/1022969.1022973. 11, 17

[202] Y. Yetim, S. Malik, and M. Martonosi, "EPROF: An energy/performance/reliability optimization framework for streaming applications," in *Proceedings of the Asia and South Pacific Design Automation Conference*, 2012, pp. 769–774. DOI: 10.1109/ASP-DAC.2012.6165058. 43

[203] M. Yuffe, E. Knoll, M. Mehalel, J. Shor, and T. Kurts, "A fully integrated multi-CPU, GPU and memory controller 32nm processor," in *Proceedings of the IEEE International Solid-State Circuits Conference*, 2011, pp. 264–266. DOI: 10.1109/ISSCC.2011.5746311. 46

[204] J. Zebchuk, V. Srinivasan, M. K. Qureshi, and A. Moshovos, "A tagless coherence directory," in *Proceedings of the IEEE/ACM Annual International Symposium on Microarchitecture*, Dec. 2009, pp. 423–434. DOI: 10.1145/1669112.1669166. 55

[205] C. Zhang, F. Vahid, and W. Najjar, "A highly configurable cache architecture for embedded systems," in *Proceedings of the International Symposium on Computer Architecture*, Jun. 2003, pp. 136–146. DOI: 10.1145/871656.859635. 52

[206] C. Zhang, F. Vahid, J. Yang, and W. Najjar, "A way-halting cache for low-energy high-performance systems," *ACM Transactions on Architecture and Code Optimization*, vol. 2, no. 1, pp. 34–54, Mar. 2005. DOI: 10.1145/1061267.1061270. 51

[207] L. Zhong and N. K. Jha, "Energy efficiency of handheld computer interfaces: Limits, characterization and practice," in *Proceedings of the International Conference on Mobile Systems, Applications, and Services*, 2005, pp. 247–260. DOI: 10.1145/1067170.1067197. 1

Authors' Biographies

MAGNUS SJÄLANDER

Magnus Själander is a Research Associate at Uppsala University. He received both his Ph.D. degree (2008) and Lic.Eng. degree (2006) in Computer Engineering from Chalmers University of Technology, Sweden. He has been a visiting researcher at NXP Semiconductors, worked at Aeroflex Gaisler, been a post-doctoral researcher at Chalmers University of Technology, and a research scientist at Florida State University. Själander's research interests include energy-efficient computing, high-performance and low-power digital circuits, micro-architecture and memory-system design, and hardware-software interaction.

MARGARET MARTONOSI

Margaret Martonosi is the Hugh Trumbull Adams '35 Professor of Computer Science at Princeton University, where she has been on the faculty since 1994. She also holds an affiliated faculty appointment in Princeton's Electrical Engineering Department. Martonosi's research interests are in computer architecture and mobile computing, with particular focus on power-efficient systems. Her work has included the development of the Wattch power modeling tool and the Princeton ZebraNet mobile sensor network project for the design and real-world deployment of zebra tracking collars in Kenya. Her current research focuses on hardware-software interface approaches to manage heterogeneous parallelism and power-performance tradeoffs in systems ranging from smartphones to chip multiprocessors to large-scale data centers. Martonosi is a Fellow of both IEEE and ACM. She was the 2013 recipient of the Anita Borg Institute Technical Leadership Award. She has also received the 2013 NCWIT Undergraduate Research Mentoring Award and the 2010 Princeton University Graduate Mentoring Award.

STEFANOS KAXIRAS

Stefanos Kaxiras is a full professor at Uppsala University, Sweden. He holds a Ph.D. degree in Computer Science from the University of Wisconsin. In 1998, he joined the Computing Sciences Center at Bell Labs (Lucent) and later Agere Systems. In 2003 he joined the faculty of the ECE Department of the University of Patras, Greece and in 2010 became a full professor at Uppsala University, Sweden. Kaxiras' research interests are in the areas of memory systems, and multiprocessor/multicore systems, with a focus on power efficiency. He has co-authored more than 100 research papers and 13 US patents, participated in five major European research projects, and currently receives funding from Sweden's business incubator and innovation agency VINNOVA. Kaxiras is a Distinguished ACM Scientist and IEEE member.

Printed in the United States
by Baker & Taylor Publisher Services